Simulating Innovation

Simulating Innovation

Computer-based Tools for Rethinking Innovation

Christopher Watts

Ludwig-Maximilians University, Munich, Germany

Nigel Gilbert

University of Surrey, UK

Edward Elgar

Cheltenham, UK • Northampton, MA, USA

Published by
Edward Elgar Publishing Limited
The Lypiatts
15 Lansdown Road
Cheltenham
Glos GL50 2JA
UK

Edward Elgar Publishing, Inc.
William Pratt House
9 Dewey Court
Northampton
Massachusetts 01060
USA

A catalogue record for this book
is available from the British Library

Library of Congress Control Number: 2013949790

This book is available electronically in the ElgarOnline.com
Economics Subject Collection, E-ISBN 978 1 78347 253 6

ISBN 978 1 84980 160 7 (cased)

Typeset by Servis Filmsetting Ltd, Stockport, Cheshire
Printed and bound in Great Britain by T.J. International Ltd, Padstow

Contents

Preface

This book was written to help researchers in the fields of innovation studies, economics, organisation studies, sociology of science and policy modelling to become more familiar with a research approach that could complement their own. Those with no background in simulation modelling may see the advantages of working with social simulation modellers, or becoming a modeller themselves. Experienced modellers will find plenty of examples of social simulation, and especially of agent-based modelling to inspire their own work. The work here is inter-disciplinary, connecting with sociology, economics, business studies and operational research in particular, but may also interest researchers into complex adaptive systems more generally, who today may be found in such disciplines as physics, mathematics and biology.

In line with the various readerships there are multiple ways to read this book, depending on one's degree of interest in the areas of application, and on one's level of technical skill.

Researchers familiar with fields such as innovation studies or organisation studies may compare the ideas expressed here using simulation models with those from other sources, especially empirical studies. All the simulations discussed in the book use our own programs, some replications of models by other authors and some models of our own design. Readers can download programs from a website (http://www.simian. ac.uk/resources/models/simulating-innovation and see appendix), gain hands-on experience of using them, and explore the behaviours of these models beyond what is discussed in the text, perhaps leading to new findings. Those more experienced in computer programming, or wishing to become so, may want to examine the code, and may be able to suggest improvements or extensions.

The best test of simulation modelling is whether someone can replicate a model and its behaviour given just a written description of it. This task, it has long been recognised, is valuable but notoriously difficult (Axelrod, 1997a, Appendix; Axtell et al., 1996; Hales, Rouchier and Edmonds, 2003; Rouchier et al., 2008; Wilensky and Rand, 2007). Example attempts are both scarce in the literature and also mixed in their degree of success (Bigbee, Cioffi-Revilla and Luke, 2007; Edmonds and Hales, 2003; Macy and Sato, 2010; Rouchier, 2003). Even when plenty of technical details have been supplied, there is still scope for the unintentional omission of vital details. In addition, attempts to implement the same model in two different programming languages or with different hardware or operating systems can occasionally generate unexpected variation in program behaviour. The threat of such problems can be reduced if attempts are made to replicate as many as possible of other authors' models. Model replication attempts are also a good way to test one's understanding of the original descriptions of the models, and in this respect the collection of programs on our website may serve as a contribution to the field.

We encourage others to have a go at replicating our models. The more successful replications are reported, the more confidence people will have in social simulation as a research tool. Hence we outline our programs' workings, aiming at supplying enough detail to give modellers some idea of their core mechanisms, without overloading the text with technical details. The World Wide Web means that our programs can easily be made publicly available to those wishing to read them.

Given the space restrictions on this book, we have chosen to focus on simulation modelling. To have attempted to review the innovation literature would either have extended the book considerably or have risked being too brief to give a fair account. Fortunately, the areas we address with our models are mostly well served for books and journals. The classic by Rogers (2003), and recent collections edited by Malerba and Brusoni (2007) and Fagerberg, Mowery and Nelson (2005) are good starting points.

As well as declining to review the literature in innovation studies, we also omit giving a basic introduction to social simulation. This has been done elsewhere (Gilbert, 2008; Gilbert and Troitzsch, 2005). There are also software packages and training materials on the Internet. Some of the programs were written using *Microsoft Excel 2003/XP with VBA*, and versions of this software are widely used in companies and universities, although problems may be encountered with other versions of Excel and with spreadsheet programs that attempt compatibility with Excel. Most of the programs, however, were written for the agent-based modelling language, *NetLogo 5.0* (Wilensky, 1999). This is free to download (http://ccl. northwestern.edu/netlogo/) and install on Windows, Macintosh or Linux platforms. We find it easy to use and relatively easy to learn to program in, it having been developed from a programming language, *Logo*, that was originally intended for use by American primary school children. One of us (NG) has employed NetLogo for research, consultancy and teaching purposes for several years, and we have no hesitation in using it here. Programmers proficient in other languages should have few problems in transferring to it, though they might prefer to try to replicate the models in their favourite language, using our NetLogo code as a guide.

One of us (CW) was introduced to agent-based modelling and the study of complex systems through reading Robert Axelrod's book *The Complexity of Cooperation* (1997a) and Stuart Kauffman's *At Home in the Universe* (1996), before attempting to reproduce the models and computer experiments described therein using what was then a fairly rudimentary knowledge of Excel and VBA. Agent-based models can be more sophisticated today, and there are now plenty of academic departments around the world specialising in complexity research, but all of the programs described in this book can be run on home PCs and we hope the subject remains one an amateur or a visitor from another field can get into. We dare not hope to have attained the heights of these two books, or go on to have the same influence, but if the next generation of modellers and complexity scientists feels inspired to apply simulation modelling when investigating issues of innovation, we shall feel this book was worthwhile.

Acknowledgements

This book is based on work conducted as one part of the *SIMIAN* project at the University of Surrey, England. The project was funded by the National Centre for Research Methods, part of the UK's Economic and Social Research Council (ESRC). Thanks must go to Edmund Chattoe-Brown, University of Leicester, who co-authored the original funding bid, and to our colleagues in the Centre for Research in Social Simulation (CRESS) at Surrey University for contributing to a rich research environment during the project. In particular, Richa Sabharwal created the *SIMIAN* website, Lu Yang provided smooth administration of the project and its events, and Lynne Hamill, Ozge Dilaver-Kalkan and Jen Badham offered fruitful conversations on the topics of this book. CW would like to thank his former colleagues at Warwick Business School, where his PhD supervisor Stewart Robinson gave him both training in the art of simulation modelling, and also enough flexibility over time to be able to pursue wider interests, some of which prepared the way for the work for this book. CW also benefitted from attending the 2009 *Tenth Trento Summer School* at the University of Trento, Italy, the 2009 *Modelling Science* workshop at the Virtual Knowledge Studio, University of Amsterdam and the 2011 *SKIN* workshop at the University of Koblenz-Landau, Germany. We thank the organisers of these events for their invitations. We also thank CW's wife, Annelise, for the motivation provided by threatening not to marry him if the book's first draft was unfinished, and for relenting when he just fell short.

Material for this book was presented at various conferences, in particular the *European Association for the Study of Science and Technology* (EASST) 2010 conference, the *Dynamics of Institutions and Markets in Europe* (DIME) Final Conference in 2011 and the *European Conference on Complex Systems* (ECCS) in 2011, and we thank the audiences on those occasions for their comments. One of the models developed for Chapter 5 has already been described in a paper in the journal *Scientometrics* (Watts and Gilbert, 2011). Chapter 6 is a slightly modified form of a paper to the DIME Final Conference. Various models by other authors have been reproduced and discussed in this book, and the references to the original sources for these are contained in the text. As we are only too aware, innovative research involves re-combinations and re-interpretations of pre-existing parts. We hope we have done justice to these various sources of assistance, and any errors should, of course, be attributed first to us.

1. Why simulate innovation?

This book seeks to innovate in the tools we use for thinking about innovation. Computer simulation models can clarify our thoughts and explore their implications. Over the last two decades there have appeared descriptions of computer simulation models that address some of the issues surrounding innovative ideas, practices and technology, including how innovations can be generated, how they diffuse among people and organizations, and the impact innovations have on people's and organizations' other ideas, practices and technologies. This book will provide a critical survey of some of these tools for thinking, while also introducing a few tools of our own.

In this chapter we explain why one might want to be thinking about innovation, how it involves complex adaptive systems and how these can be studied, and hence why one might want to add computer simulation models to the tools one uses for innovation studies. The chapter concludes with an outline of the rest of the book.

WHY STUDY INNOVATION TODAY?

The Trouble with Financial Innovation

Innovation is currently held responsible for a lot. During the work for this book (2009–12) countries around the world have been suffering the after-effects of a wave of innovation in the financial world. The tale, as told by *Financial Times* journalist and trained social anthropologist Gillian Tett (2009), tells of brilliant minds being hired by investment banks and, full of excitement for their work, putting in long hours to generate innovative ways of making money (see also MacKenzie, 2009, 2011a). They began with the idea of extending the centuries-old concept of derivatives, a form of insurance, to a new application, that of insuring against the risk of a borrower defaulting on their debts. The tale is woven around a diverse collection of novel financial concepts and products, each requiring a new name or phrase: from CDS to CDO, CDO of ABS, mortgage-backed CDO, slice-and-dice, tranches, CDO-squared, Gaussian copula, sub-prime, super-senior, SIV and ABCP. Each one represented a new combination of pre-existing components or a new application for an existing idea. Once invented, the innovations were offered to new markets, scaled up to new levels, and sold off in unprecedented quantities to a variety of customers. These customers included not only traditional, but also some new financial players, most of whom had little knowledge about how these financial products worked. Understanding of the risks and value behind these novel products was also scarce among their producers, their owners, government regulators and insurers, but few saw any incentive for asking questions. When the underlying assets, mostly mortgages on houses, began to lose their market value, a house of cards was set to tumble, freezing markets and taking down banks and brokerages, insurers, investment and hedge funds, government finances and even governments themselves, and finally reduced the power of several nations. All this followed a frenzy of innovation.

At the same time, innovation is at the heart of proposed solutions to the crisis. Governments should introduce new regulations for the banking sector. R&D spending should be increased in other industries, especially manufacturing, in order to generate new growth to compensate for the losses. The gap between universities and businesses must be narrowed, with exchanges of knowledge between them, and more patents generated and spin-off companies set up based on academic research. Geographic clusters of firms must be seeded and protected, where interactions between the firms will generate the next big ideas in technologies. So innovation has been hero, then villain, and is now our best hope for salvation. It seems an apt moment to be writing about innovation.

The Trouble with Economics

Given the economic causes and effects of innovations such as those in the financial world, it might be thought that the topic of innovation would best be studied by economists. The primary focus of mainstream economics is efficient resource allocation, for which mathematical models have been developed based on the idea of a system in equilibrium. Solow (1956) provided a mathematical treatment to add resource growth to modelling as part of a dynamical equilibrium theory, but these models assume both population growth and technological change are givens, exogenous to the model. By this light, technological innovation is just an unexplained leftover when one has subtracted other factors behind resource stocks. Endogenous growth theory (e.g. Romer, 1986, 1990) considers some of the factors thought to be behind technological change, chiefly those that increase human capital, knowledge and innovation, such as R&D

spending, the level of government regulation and a culture of openness to change. A key difference from previous economic theories is the idea that investing in R&D can produce *increasing returns* to scale: acquired knowledge enables improvements in future knowledge production.

These attempts to study innovation endorse most of the common assumptions of mainstream economics, such as rational agents forming systems at equilibrium, and largely consist in developing equation-based models that will reproduce statistical patterns observed in data, in this case, data by country on GDP and growth, population size and R&D spending, among other measures. In so far as correlations are found between these variables, how the correlations come to be there is poorly understood. Representing the generating mechanisms means representing human behaviour, including representing its diversity, mathematically in such a way that it can be aggregated easily.

While mainstream economics remains attached to its assumptions and mathematical techniques, it continues to treat the topic of innovation poorly. This can be seen by the continued neglect in mainstream economics textbooks of fields that deal primarily with innovation: evolutionary economics and behavioural economics.

The pioneer of *evolutionary economics*, Joseph Schumpeter (1939, 1943), writing in the middle of the twentieth century, identified innovations and the entrepreneurs who develop innovative ideas into marketable products as vital to economic growth (Heilbroner, 2000, Chapter 10). The theories of neoclassical economics focus on markets at equilibrium. But according to these theories, at the equilibrium point, competition between firms has reduced profit to zero. In this case, why remain in the market? This seeming puzzle could be solved, according to Schumpeter, by reference to innovation. When companies bring new products to market, or develop improved methods of production resulting in lower costs, they enjoy an advantage over their competitors and may charge prices that include a premium, thus yielding non-zero profits. Their new offerings may also enhance the value of other goods and services, and undermine the market appeal of yet others, a process Schumpeter dubbed 'creative destruction' (Schumpeter, 1943, p. 83). The advantage is only temporary, however, since competitors may imitate the innovator. For this reason, some of the innovator's profit should be invested in the R&D that could generate future innovations and maintain some competitive advantage. Alongside efficient allocation of resources, forcing firms to innovate is the second major justification for markets. But uncertainty exists about how much to invest in this R&D, how best to go about seeking innovations, how much one innovation depends on knowledge of another and how long it will take to generate the next one. Different companies may adopt different strategies for this, with some investing heavily in R&D and others hoping to be able to imitate quickly and cheaply when the investments by others have generated results. At some times there may be a flurry of new products, at other times the diffusion of recently introduced products, and at yet other times there may be a period of relative quiet, perhaps resembling a market equilibrium state. Thus, while undergraduate economics courses teach students to focus on the ideas of equilibrium being reached by a market of identical competing firms, the vision developed from Schumpeter's work is that of *heterogeneous* (diverse) firms in a *dynamic* market.

Another field trying to attract more attention within economics is *behavioural economics*. When reasoning about the decisions made by suppliers and customers, neoclassical economics assumes that decision makers know all the available options, the probabilities and monetary values of all consequences of these options, and will choose between the options so as to maximise their expected monetary gain. This view of human decision makers as rational optimisers with perfect information, or *homo economicus*, was criticised by the political scientist, Herbert Simon, beginning in the 1940s and continuing in the decades since (Simon, 1948, 1955a, 1957, 1991). In its place, Simon and collaborators proposed that human decision makers had limited information on options, probabilities and values, and limited ability to process the information they had in a short enough time for it to be useful. Instead, of being infinitely capable rational optimisers, 'bounded rational' humans employed relatively quick and easy rules of thumb, called heuristics, to search for solutions that were, if not the best possible, usually sufficiently good for survival (Simon, 1955a; Simon and Newell, 1958). Nelson and Winter (1982) combined this view of bounded rational agency with evolutionary economics. Laboratory experiments by psychologists Kahneman, Slovic and Tversky (1982) confirmed that how human beings actually performed decision making resembled the use of heuristics more than it did mathematical optimisation. Both Simon and Kahneman have since been rewarded with Nobel Memorial Prizes in Economic Sciences (in 1978 and 2002, respectively). In the 1990s, support for research into actual economic behaviour continued to build (Akerlof and Shiller, 2009; Kahneman, 2011; Klein, 1998). More recently, interest has grown in the study of what it is that decision makers seek to improve, in particular, happiness (Frey, 2008; Layard, 2011), instead of money. Despite this, an informal survey of the undergraduate-level textbooks in the economics sections of bookshops and libraries reveals that most still lack chapters devoted to either evolutionary or behavioural economics.

Following the financial crisis, however, confidence in mainstream economics has been shaken (Blanchard, 2012; Frydman and Goldberg, 2011;

Keen, 2011; Turner, 2012). There is an opportunity for rethinking the subject's core material, that is, what is taught to students, and also what is funded, what research is published in the most widely read journals, who gets employed by the most prestigious academic institutions and who will go on to influence the next generation of society's leaders. Time and effort is being devoted to innovative approaches, be these either the invention of new methods, or the importing of ideas from other fields, including psychology, sociology, neuroscience, cognitive science, biology and the various fields which study complex adaptive systems.

New Sources on Innovation

The information age has brought new data sources to help the change in focus. There is more emphasis on attempts to count innovations. In technology there are data on patents, including who patents what, who they patent it with and which patents refer to which others (Fleming, Mingo and Chen, 2007; Fleming and Sorenson, 2001; Sorenson, Rivkin and Fleming, 2006; Trajtenberg, 1990). Similarly, data on academic publications, their co-authors and their citations, give insights into innovation production within universities and other research institutes (Boerner, Maru and Goldstone, 2004; Goldenberg et al., 2010; Price, 1965; Small, 1973). Electronic records of individuals' interactions, such as email communications, the Internet Movie Database (www.imdb.com) or geographical tracking devices can provide impressions of the social networks within which information about innovations flows and ideas are combined to generate new innovations.

In addition to these quantitative sources of data, qualitative sources, especially ethnographic studies over the last 30–40 years, have caused a revision of views of innovation generation and adoption. Seen close up, the supposed events of invention and adoption of new ideas, practices or products become more complex and less identifiable (Akrich, Callon and Latour, 2002; Akrich et al., 2002; Bijker, 1995; Bijker and Law, 1992). Since the 1990s, developments in artificial intelligence, robotics and cognitive science (Clark, 1997; Hutchins, 1995a, b) have promoted a view of the human decision maker as being *embedded, embodied* and *social*, with decisions dependent on a historical context, on interaction with a material environment and on collective effort.

It remains to be seen whether analyses of these quantitative and qualitative datasets will lead to better policies on innovation. Some uses of the datasets, such as policies that attempt to base continuance of funding on past production of patents or publications, could cause innovators to adapt their behaviour from that which helped generate the past data.

Unlike, say, astrophysicists, social scientists have the potential to disturb the systems they study. However, where policy and behaviour has yet to reflect the results of analyses, the datasets may help us to understand retrospectively how innovations were generated, how they interrelate, how they diffuse and what their impact may be.

Both quantitative and qualitative studies can inform the creation and revision of theories about innovation, which in turn can inform policy making. Theorising, however, can be hard to perform in unambiguous, coherent detail, with its implications spelled out. The time is ripe for a technique that allows theorising to capture some of the complex networks of interdependencies, and the dynamic behaviour that results. In recent decades a new type of tool has emerged for improving the rigour of theories and exploring their coherence and consequences, generating new hypotheses for empirical studies (Davis, Eisenhardt and Bingham, 2007). These are computer simulation models, and this book applies them to the study of innovation. In this we draw upon papers and books by others that have appeared over the last 20 or so years. These works apply simulation models to the diffusion of innovations through social networks, to collective learning in organisations, to the structure of academic science publications, to the adoption and adaptation of technologies in complex contexts and to technological evolution and the formation of innovation networks, to name the major topics of our chapters. Given that innovation remains as important an area as ever, and given the numbers of these tools, it seems a good time to highlight some of the models, including their features, assumptions and purposes, and identify some recommendations for future models.

WHAT IS MEANT BY 'INNOVATION'?

A Few Common Distinctions

There are many uses for the word 'innovation', and uses in this book will reflect several different bodies of literature, although the authors of models can be quite vague about the types of innovation they intend to apply them to. A few common distinctions may be made, however.

Two ideas seem essential to the concept of innovation. The more obvious idea is that it involves *newness*, or novelty. For example, there may be a new item or service brought to market (*product innovation*), or a new method for producing a product more cheaply than before (*process innovation*). The second idea is that the new thing will be of some value to someone, that is, it will be an improvement, reaching a new level of quality, useful, relevant or appropriate to the concerns of some person or persons, be they customers in a market, academics in a scientific field or workers in a firm. Combining, for example, a banana with a spreadsheet may be an original or novel idea, but if it has no use or makes no sense to anybody, it seems pointless to treat it as an innovation. This does not mean bananas and spreadsheets will always remain apart, however. We cannot rule out the possibility that someone, somewhere, will one day find some meaning in that particular combination. Indeed, it may be that someone already has, but that the news of this has yet to reach us. It would seem, then, that both ideas implicit in innovation, novelty and appropriateness, are relative to some particular audience. Like beauty, innovation is in the eye of the beholder.

We will write about 'innovation' as a process and about 'innovations', the objects of the process. In fact, there are two types of process going by the name of innovation. The first is the *generation* of an innovation, for example, the combining of two ideas to make a third idea. The second is the introduction of an innovation to a group or a market, and its spread thereafter, a topic usually called the *diffusion of innovation*. However, both generation and diffusion will be represented in most of our models.

There is a formal model of innovation taught in business schools called the *linear model*, which focuses attention on a division into separate stages, innovation generation followed by innovation diffusion (Godin, 2006). Sometimes the first phase is divided further, between the results of basic research, such as a prototype or a discovery, and the results of development of the research results into an application or marketable product (e.g. Cooper, 1990). In recent decades, these divisions have received much criticism (Balconi, Brusoni and Orsenigo, 2010). Some of the models in this book will represent the distinction, others will reflect the critiques.

Part of the grounds for dispute is that an innovation can be both a physical construction, and also something less tangible: an idea, practice or meaning. The former might be, for example, an object produced by the particular combination of components which have never before been combined in this way (*recombinant innovation*). This novel object, however, may then be used for an old, familiar purpose. Conversely, an existing, familiar object may be used in a novel way, and given a new interpretation or value (*transfer innovation*). If the object is then modified physically in order to improve its ability to enact the new application, is this an additional innovation? Likewise, if our behaviour adapts to a new object because the old purpose is only imperfectly served by it, should we count this an additional innovation? Again, the models in this book will vary in where they identify innovations.

Innovations, whether products or processes, have relations to each other. Some new things can replicate, partially or wholly, the functions of others. These can then serve as *substitutes* for each other. Other new things enhance the functionality and value of existing products and processes. These are *complementary*. In some cases, the effect of one item on the ability to produce or use another may be quite strong. If product B does not work without the use of product A, product A having no substitutes, then A is *necessary* for B. If B is an automatic consequence of A being used, with no other products necessary, A is *sufficient* for B. *Dependency relations* between things will play an important role in innovation studies.

Several typologies found in the innovation literature are worth mentioning. First, a distinction is sometimes drawn between *qualitative* and *quantitative* innovations. The invention of the aeroplane, combining the internal combustion engine from automobiles with the wings-and-tail airframes of gliders, is a qualitative innovation. It produced capabilities very different to those of the car or the glider, including airborne reconnaissance, bombers and fighter planes. In comparison with ships and trains, however, it offers primarily a quantitative improvement: faster transport of people. Whether one sees an innovation as quantitative or qualitative depends on which dimensions one focuses on and which comparison technologies one selects.

Other distinctions to be drawn are those between *radical, incremental, modular* and *architectural* innovations (Henderson and Clark, 1990). Consider an electric air fan, mounted on the ceiling of a room. Improvements to the blades or the motor would be incremental. Replacing it with another technology, such as air conditioning – based on a very different principle and physical phenomenon but aimed at similar effects – would be radical. The fan blades and motor could be reassembled in another way, as a desk-mounted portable fan. Such rearrangement of component parts would be architectural innovation. Keeping the configuration of components, but replacing one of them with a new technology – such as a new type of motor to replace the electric one – would be a modular innovation. Incremental innovation maintains both the core design concepts of a technology and the linkages between its concepts and components. Modular innovation involves a change in the core concepts. Architectural innovation involves a change in the linkages. Radical innovation involves both. Henderson and Clark, who introduced this framework, admit that 'the distinctions between radical, incremental and architectural innovations are matters of degree' (Henderson and Clark, 1990, p. 13). But they invoke the framework to explain the relative degrees of disruption that technological innovations can cause. If a producer firm is well-established, incremental changes build upon its core competences and have little effect on its strategy and organisation. Radical innovations render its expertise obsolete and can be devastating to the way it runs itself. Architectural innovation, however, mostly preserves the usefulness of the firm's knowledge of the components, while demanding that the firm rethinks how it uses the components. The combination of giving up some areas of expertise while preserving others may prove difficult for an established firm – in contrast, perhaps, to a young start-up firm with no emotional attachments to the expertise of particular staff members, and no financial investments in particular production machinery. A modular firm, that mirrors the structure of its product technology by its organisational structure, may find it easier to handle modular and architectural technological innovations as changes respectively within and between organisational units. Thus technological and social organisational structures can interact.

Creativity Myths and Some Insights into Innovation

As human beings we love to tell stories, including stories about inventions, discoveries and how some important component of our present lives came into being. But what makes for a good story may not reflect the real processes of innovation. Sawyer (2012) notes that Western societies employ a number of 'creativity myths' concerning how the fruits of creativity, or innovations, come about. Inventions are given mysterious origins: the flash of genius, the bolt from the blue or divine inspiration. We also like to think of creation being the act of a lone individual: the misunderstood artist starving in his garret, producing masterpieces that will not sell until he is dead; or the mad scientist, living in obscurity, and making the discoveries that will go unrecognised for 50 years. Even our legal systems reflect a focus on individuality, with patented inventions giving special rights to the holders, who may or may not have been the first to have the patented idea, and who may or may not be capable of developing a commercially viable product from it.

The reality may be more complicated. Sawyer lists a number of insights into business creativity (Sawyer, 2012, p. 285), which he illustrates with the case of the development of the graphical user interface (GUI) of windows, icons, menus and pointers (WIMP) by Xerox, Apple and Microsoft. We repeat the insights here (in italics below), but illustrated with our earlier case of credit derivatives (Tett, 2009).

Each innovation builds incrementally on a long history of prior innovations. One might wish there to be some invention or decision event, to which one could point saying, 'Ah! That was when credit derivatives were created which would destroy the financial world!' As mentioned already, the real history does not oblige this wish, with developments going back decades and even centuries. Concerning the GUI, the elements of the WIMP paradigm came together from several sources and several stages, including academic research in the 1960s, demonstrations of the personal computer concept by Xerox in the 1970s and computers intended for mass production launched in the mid-1980s. The market dominance of Microsoft's *Windows*, today the best selling operating system, obscures the fact that it took major revisions before a PC-compatible GUI achieved commercial success with *Windows 3.0* and *3.1* in the early 1990s and another major revision, *Windows 95*, was needed to attain the GUI standards set by mid-1980s computer platforms.

Innovations emerge from collaborative teams. Tett describes a collection of brilliant individuals with international backgrounds and some diversity in education and career paths. Among them were people with strengths in banking itself, mathematical modelling and customer relations. Beyond this team there were lawyers and technical services, representatives of the customers, and regulators, all of whom had some input into the development of the new types of deals. If the development of an innovation involves multiple insights, each of these probably comes from a different team member. But a single person usually comes to be associated with the innovation, a Thomas-Edison type, who then gains the most in reputation from it. In the case of credit derivatives, one member of the J. P. Morgan team became a spokesperson for the new field, with many media appearances, and reportedly has received hate mail since the financial crisis emerged. However, as Tett points out, the team at J. P. Morgan in no way intended or anticipated how their idea would be reapplied at other banks, and at some stages were alarmed at the growth in scale of the new markets.

Multiple discovery is common. Tett focuses on one team at J. P. Morgan, who have some claim to having been the first to arrange a credit default swap deal. But rival firms were quick to copy this type of deal, partly because they were already familiar with the concepts involved, and competition during the growth years of the credit derivatives markets was intense. In the case of the multiple GUI developers, the extent to which they acted independently rather than imitated was a matter for legal disputes.

There is frequent interaction between the teams. In the case of the GUI, both Apple's Steve Jobs and Microsoft's Bill Gates visited Xerox PARC and saw some of their developments. In the case of banking services, staff are often poached by rival firms, sometimes taking whole groups of colleagues with them. In addition, mergers between firms led to rival teams being brought together.

A product's success depends on broad contextual factors. In the case of the GUI, the cost of a personal computer employing the new software designs was prohibitive until the late 1980s. It took the spread of the idea of desktop publishing, a so-called 'killer app', to boost sales of Apple's Macintosh computer. The growth of the credit derivatives markets was helped by a number of contextual factors. To fuel it, there was a plentiful inflow of money from pension funds, oil sheiks, a policy of cheap lending from the US Federal Reserve, and a growing trade deficit with China. From the 1980s on there was a celebration, especially on the part of politicians, of the entrepreneurial work performed in financial markets, including the invention of new product types, the adoption of new ways to trade, such as computer-based trading, and acceptance of new levels of risk taking in order to seek out the best returns for one's clients and shareholders. These cultural factors led to the removal of old regulations, the removal of restrictions on transactions, especially cross-border ones, and a relaxed, *laissez-faire* approach on the part of governments to the introduction of regulations to deal with the new financial products. The culture also infected people outside the financial industry, such as local government financial operators, who sought riskier places to invest money on the grounds that it would maximise returns – not previously thought of as an aim of local government. Even when concerns were expressed about the new markets, especially their rapid growth and sheer scale and the lack of accurate assessments of these, the growing practice of paying lobbyists to represent the interests of businesses to politicians meant that attempts to regulate the financial sector more closely were headed off or watered down. Furthermore, the novelty of the products obscured what was familiar. While some might have identified the boom as an old-fashioned market bubble, and plenty have since noted the parallels with the Wall Street Crash and the resultant Great Depression, at the time the novelty of the products meant that, as with the Dot.Com boom in the 1990s, people could argue that 'this time is different'. So, rather than one person, one team or one bank being responsible for the products behind the financial crisis, we find whole societies collectively creating the situation they then find themselves in.

Innovations, then, despite the creativity myths, seem to be the result of collective labours from many participants, in many locations, taking many steps, involving many components with many connections between them. Innovations emerge from the interactions of a complex system of social and technological parts. If we are to think about innovation then we need tools for dealing with complex systems and the emergence of novelties from these.

INNOVATION AND COMPLEX SYSTEMS

The Growing Interest in Studies of 'Complexity'

Having suggested some reasons for studying innovation and what that might involve, we turn to the second theme of this introductory chapter, and of the book itself: complexity, emergence and complex adaptive systems.

An innovative academic field has emerged over the last few decades with complex systems as its focus. It is best encapsulated by the Santa Fe Institute (SFI), founded in the 1980s and still serving as an inspiration for researchers (www.santafe.edu/about/history). Part of this influence stemmed from the publication of engagingly written early accounts of its work by popular science writers (Lewin, 1993; Waldrop, 1993), but from early on, the subject matter was engaging as well. Unlike most funded research projects of the time it did not engage in providing confirmation of small-scale, pre-specified beliefs within established academic disciplines, but instead aimed at bigger, more fundamental questions, and transcended disciplinary boundaries. SFI brought together established researchers from multiple disciplines, beginning with physicists from nearby Los Alamos, and adding economists when Citicorp, keen to develop alternatives to mainstream economic thinking, offered to fund some economists known for their more maverick interests in technology and evolution. Biologists, mathematicians, anthropologists and computer scientists followed, as well as cognitive scientists and psychologists in the 1990s, and the occasional artist, especially writers. The biological sciences made particularly important contributions, including theoretical biologists debating evolution and adaptation, the origin of life as self-organising systems, the emergence of cooperation, and the population dynamics of ecosystems, where complex networks of who-eats-whom relations made different species interdependent for their evolutionary success in complicated, impossible-to-predict ways.

A common thread in SFI work throughout its history was the development of new computer tools for studying complex systems, including agent-based simulation models (which we will return to in a later section), statistical data analysis and pattern-recognition tools, and problem-solving heuristic search methods, such as genetic algorithms. Many SFI researchers believed that they each faced analogous problems which might be tackled using ideas and techniques inspired by each other's disciplines. Once SFI acquired buildings, visiting scholars from all over the world and from multiple disciplines were able to share offices or meet in corridors and the canteen, and discuss ideas for each other's problems.

A common feature of these problems was that they involved systems composed of multiple interdependent parts, and the behaviour of the system could not be equated to a linear sum of the behaviours of the parts. Unlike Newtonian mechanics, where vector sums are made to calculate the overall behaviour of some combination of forces, mainstream econometrics, where market behaviour was assumed to be an aggregation of individual customers and suppliers, or social statistics employing linear regression models, in these complex systems behaviour was assumed to be *nonlinear*. The scale of the effects was not proportional to the scale of the causes. For instance, doubling the person-months invested in a project might not double the output or halve the project duration. More people mean more interdependencies between them, and more time means more opportunity for adapting to each other and the task. The overall effects of additional resources may difficult to anticipate. Experience of software development taught that adding manpower to a late project makes it later (Brooks, 1975).

Also common to these problems was that the systems involved were dynamic, or changing, with component parts continually adapting to each other, in contrast to mainstream economics' focus on systems at rest, or equilibrium. Among the phenomena that could be found in nonlinear, complex adaptive systems' behaviours were phase transitions, sudden shifts in the nature of the system, in response to perhaps only small changes in a single parameter. Phase transitions were familiar to physicists from the study of matter, for example where gradual increments in temperature can cause ice to melt quite suddenly to a liquid around 0°C, and boil quite suddenly around 100°C. Another key concept was emergence, the surprising appearance of some kind of order or pattern where previously there had been only disorder. The best known phase transitions were those between order and disorder, or 'chaos' as it was popularly known, though no real connection was ever established from complexity to chaos theory, made famous slightly earlier (Gleick, 1987). Systems frozen in ordered states had no interesting consequences; neither did those in random flux. Interesting phenomena in nature and social systems could be found in-between, 'at the edge of chaos' (Lewin, 1993; Waldrop, 1993). This evocative phrase quickly spread to, among other places, the business literature (Cohen, 1997; Conner, 1998; Pascale, Millemann and Gioja, 2000) where the new 'science' was mined for metaphors for how organisations should be run: not planned and controlled by management at the top – businesses and economies as complex adaptive systems were too unpredictable for that – but instead directed from the bottom up, with workers and other system components given sufficient freedom for the business to self-organise and its new policies to emerge.

After a wave of publicity, SFI encountered scepticism about whether a general theory of complex systems was a sensible goal – complex social systems might have very different laws to complex biological systems, for example – and whether any contributions recognisable to other scientists would emerge amid all the dreams and hype (Horgan, 1995). In response, SFI researchers began to rely less on toy models of abstract complex systems, and more on empirical studies. In this they were helped by work from elsewhere that raised awareness of the importance of network structures in biological and social systems (Barabási, 2002; Buchanan, 2002). New technologies meant that large-scale datasets were being generated in biological and social sciences that could be analysed to test hypotheses about how the real systems functioned.

The Multiple Discoveries of Complex Systems

It should be noted that the researchers meeting at the SFI were neither the first nor the only ones to think about complex systems. Bronk (2009), for example, argues it is possible to identify many of the complexity ideas relevant to economists in the works of the Romantic philosophers of the early nineteenth century. Even if attention is restricted to the second half of the twentieth century, there are still plenty of examples, of which we can cite a few.

The work of Herbert Simon and his collaborators from the 1940s onwards at what is today Carnegie-Mellon University has already been mentioned. Simon's paper on 'The architecture of complexity' (Simon, 1962) was an early contribution, arguably decades before its time. His former co-workers pioneered the computer simulation of organisations, including how individual workers could collaborate to learn more than they could in isolation (Cyert, March and Clarkson, 1964; March, 1991). Chapter 4 will draw upon some of these ideas. This work was combined with themes from evolutionary economics (Nelson and Winter, 1982).

Also stemming from the 1940s and '50s, cybernetics considered the dynamics of systems, including business organisations, for the interdependencies between their parts, especially *feedback* and *feedforward loops* that could regulate a system or send it spiralling out of control (Ashby, 1956; Wiener, 1948). One result from cybernetics was Ashby's law of requisite variety, which holds that for a subsystem to be controlled (for example, by a manager) the variety in its behavioural states must be matched by the variety in the states of the controlling system (Ashby, 1956, 1958). The General Systems Theory of Bertalanffy (1971) attempted to apply the concept of a biological system, the organism, to the phenomena of other disciplines. Both cybernetics and systems theory owed their appeal to a belief that studies of systems in general could inform attempts to manage human systems (Beer, 1959). Systems thinking prompted the development of system dynamics modelling (Forrester, 1961; Sterman, 2000), a computer simulation of stocks and flows of materials, people or other quantifiable things, in which the behaviour of any one stock level could depend on the levels of stocks and flows elsewhere in the system. Another outcome of systems thinking was a focus on how firms adapt over time in response to experience of an environment, in the *learning organisation* (Senge, 1992).

Those schooled in the mathematical techniques of operational research (OR) or management science became increasingly concerned during the 1970s that the real-world problems to which they tried to apply their skills did not resemble textbook exercises or idealised conditions (Rosenhead and Mingers, 2001). Instead, organisations presented them with 'messes' (Ackoff, 1981), 'wicked problems' (Rittel and Webber, 1973) and 'swamp conditions' rather than the high ground (Schön, 1987). Indeed, often the biggest challenge was in identifying what problems were faced within the organisation, rather than in solving the problems. Since then expertise has been developed in so-called 'Soft OR', applicable when systems include hard-to-quantify phenomena, such as opinions, norms, politics and emotions. Techniques for facilitating group discussions and involving stakeholders are employed in order to improve members' understanding of their own organisation's situation, seek consensus as to what problems should be dealt with, and generate more buy-in for candidate solutions (Rosenhead and Mingers, 2001). Examples include soft systems methodology (Checkland, 1998; Wilson, 2001) and causal mapping/cognitive mapping (Ackermann and Eden, 2011; Eden, 1988). Outside of business organisations, those working in the field of human–environment relations, studying social-ecological systems and their sustainable development, have adopted ideas from complexity science and a belief in the value of stakeholder participation (Berkes, Colding and Folke, 2003; Voinov and Bousquet, 2010).

Working primarily in the sociology of science (to which we will return in Chapter 5), Robert Merton identified several phenomena relating to how a complex society produces and uses innovations, including 'multiple independent discoveries' (Merton, 1973, Chapters 16–17; Zuckerman, 1979), 'unintended consequences' (Merton, 1968b, Chapter 15), 'self-fulfilling prophesies' (Merton, 1968b, Chapter 16) and serendipity (Merton and Barber, 2004). Another sociologist, Luhmann, drew upon Maturana and Varela's work in biology to pursue a theory of communication and self-reference, based on their concept of auto-poietic ('self-producing') or self-organising systems (Luhmann, 1990; Maturana and Varela, 1980; Mingers, 1995).

Anthropologists had long been aware of the complexity of the social systems they studied, but the application of ethnological methods to study how scientists worked and how technologies were developed and used transformed understanding in Science and Technology Studies in the 1970s and 1980s (Bijker, Hughes and Pinch, 1987; Bijker and Law, 1992). (We will return to this in Chapter 6.) In the new picture, human agents, technologies and practices become intertwined in a complex network of social, economic, political and physical relations.

Interest in social networks began with empirical explorations of network-related social phenomena, such as the small-world effect (Killworth and Bernard, 1978; Milgram, 1967) and the strength of weak ties when seeking information on job opportunities (Granovetter, 1973), and with the development of statistical metrics for network structures in social network analysis (Boorman and White, 1976; Lorrain and White, 1971; Wasserman and Faust, 1994; White, Boorman and Breiger, 1976).

A number of physicists became interested in the phenomena of dissipative thermodynamic systems, that is, systems which take in free energy from the outside (such as from the sun) but partially constrain its dissipation, so that on a local scale (e.g. on planet Earth) ordered structures can build up and entropy decrease, in contrast to the second law of thermodynamics, which concerns closed systems and the increase of entropy. In this respect, the physicist Prigogine's work on self-organisation and irreversible systems is particularly notable, including his collaboration with the philosopher Stengers (Prigogine and Stengers, 1984). From the 1970s on, other physicists had begun to apply the techniques of statistical mechanics to social interactions (Galam, 2004), though it would take until the 1990s for this work to become well known, for example Bak's popularisation of his theory of self-organised criticality (Bak, 1997). In particular, social physicists have turned their attention to social networks, including relating network structural properties to processes of network growth and change (Albert and Barabási, 2002; Barabási, 2002; Newman, 2003, 2010; Newman, Barabási and Watts, 2006).

So, many of the ideas promoted by the SFI as part of a 'Complexity Science' or 'Complexity Theory', such as the importance in social systems of complex networks of interdependencies, were available from alternative sources. Many past researchers have been interested in transferring biological metaphors and formal or mathematical models of interdependent systems to other fields, especially those involving people. However, SFI has been a powerful inspiration for academics working in this area. Today journals exist dedicated to a complexity science approach. In the United States and UK several major universities have set up research centres in 'Complexity Science' and are now producing MSc and PhD students specialised in its techniques and topics. Papers drawing upon complexity science have appeared in mainstream journals in a variety of more traditional subject areas, with biology and statistical physics the greatest beneficiaries. However, as this book will show, social sciences can use its concepts as well.

The Variety of 'Complexities'

Both SFI-style complexity science and the fields invoking similar concepts have used the terms 'complex' and 'complexity' in diverse ways. SFI's Seth Lloyd published a list of dozens of different definitions and measures of 'complexity' (Lloyd, 2001), but he identified three emerging themes: (1) 'difficulty of description' (typically measured in bits), (2) 'difficulty of creation' (measured in time, money, energy, etc.) and (3) 'degree of organisation', subdivided into (3a) 'difficulty of describing organisational structure, whether corporate, chemical, cellular, etc.' ('effective complexity'), and (3b) 'amount of information shared between the parts of a system as the result of this organisational structure' ('mutual information'). We shall not attempt to rival this list, but instead pick out some of the main complexity concepts relevant to a book on simulating innovation.

The first point has been mentioned already: complex systems consist of multiple parts or agents. For example, many people may be involved in the production of some innovation, which then diffuses among many others. Innovations themselves may be composed of multiple components. Because of the multiple parts, recording the state of this system may take many bits of information – part of Lloyd's 'difficulty of description' theme. Each part may have multiple attributes, and each attribute may take multiple values. For example, if there are n agents in some population, and each agent has F cultural features or dimensions within which they can differ from each other, and q cultural traits within each dimension, then there are $q \wedge (n * F)$ different states of the system. If the population n increases linearly, the number of system states goes up exponentially. For quite modest values of population size, number of attributes and number of attribute values this can produce dauntingly large numbers. This is *combinatorial complexity*, the number of different ways of combining things.

A lot of problem solving can be thought of as involving a search within a vast space of possible combinations for the optimal solution, or at least a satisfactory one (combinatorial optimisation). In some cases (though not all), following a rule of thumb during this search – Simon's heuristic search methods – may be an efficient way to get from one solution to a much better one in a relatively short number of steps, which is much more appealing than trying out every possible combination in turn. If it

is necessary to try nearly every combination, this is a sign of high *computational complexity*, and introduces Lloyd's 'difficulty of creation' theme.
Our collective ability to solve combinatorial problems during our limited
lifetimes, such as innovating in our individual attributes in order to
minimise conflicts with each other and maximise mutual benefits, depends
upon the difficulty of this task and the particular methods in use for
searching through the space of possibilities.

Mainstream economics has tended to focus on collections of *homogeneous* agents, identical in their attributes and behaviour and therefore easier
to represent in mathematical calculations, perhaps as some idealised Mr
Average. This neglects the fact that real people and firms are *heterogeneous*, varied or diverse, in their attributes, not least in their spatial locations.
In addition, the attribution of a particular attribute value to an agent may
be a simplification as well. People can vary over time, often for reasons not
perceivable to them or us. Random variation over time due to chance, or
stochasticity, introduces more diversity into a system. Sometimes it does so
in well-behaved ways, ways that are regular enough to show an identifiable pattern if we collect data, plot frequency distributions and analyse statistics. These then become knowable unknowns, predictable uncertainties,
manageable risks. But when is it safe to assume we have now identified the
correct pattern? And how do we know the pattern observed in the past will
continue in the future? Even when economists recognise the possibility of
random variation, there is a temptation to assume manageability of risk,
because then they can employ mathematical techniques to deliver explanations for what has happened and advocate new policies for the future. It
can feel reassuring to hear someone claim the world's complexity can be
tamed, but they can still be wrong.

What makes the behaviour of a complex system so difficult to describe
and predict is the interdependencies between the component parts. The
systems are adaptive; parts change their states or attribute values. Reasons
for state changes include manipulation by some agent, such as a human
designer or manager, chance mutation and cross-over, such as when genes
are reproduced, and random fluctuation or noise, such as happens due to
heat. When changes occur, however, the new state is not purely a question
of chance. The probabilities of different states being adopted by one part
are determined by the states of other parts. Once a part has changed, its
state may then affect the probabilities of other parts changing. In chemistry, the presence of molecules of one compound can catalyse, or raise the
chances of reactions between other molecules. In technology the capabilities of one component are enhanced or inhibited by the other components,
and also by how people use them. In ecosystems a population belonging
to one species can affect the species it preys on and the predator species

that prey on it, as well as competitor species. Evolutionary changes in one species' behaviour and population can affect the fitness value of variations in other species, in a complex process of co-evolution.

In social groups the presence of one person can inspire or constrain the behaviour of other members, including affecting whether they continue to remain in the group. In a person's life, one behavioural practice can affect the performance of other practices, such as by learning skills that can then be transferred to other applications, and by reducing the amount of time and resources one has available for other activities. People, their beliefs and ideas, their practices, their props and tools, the places that serve as venues for them, and their social networks or knowledge of other people can all be intertwined in networks of interdependencies, catalysis and constraints. The number of such interdependency relations gives us another aspect of complexity. However, what happens in networks is not just the result of the number of links, but also other features of its structure (a point we will demonstrate in several models, and especially in Chapter 3). One important feature is the presence of loops, whereby a change in one component can initiate a chain of changes in other components, eventually affecting the original component itself. Such loops are known variously as self-referring, self-reflexivity (Popper, 1957; Soros, 1988), virtuous and vicious circles, self-fulfilling and self-refuting prophesies (Merton, 1968b), and positive and negative feedback loops (Arthur, 1994).

For example, suppose a person whose opinion is sought – a government finance minister, perhaps, or a technology consultant – makes a prediction that some tradable item, for example a company stock, will increase in price over the next year. Those who believe this person will decide they can make money by buying this stock now and selling in a year's time. Others may realise that one effect of this first group buying the stock will be to increase its price in response to the increase in demand. So even if they did not trust the first person predicting its rise, they may still believe the price will go up, and therefore that they too can make money by buying it now and selling later. This additional increase in demand again increases the price. Thus the first person appears to have made a self-fulfilling prophecy. But what if the original opinion was that the price would peak in a year's time? Those acting on such a belief will want to sell before the stock goes into a decline. They may therefore seek to sell before or at the predicted time of the peak. When they do so, this drop in demand will have an impact on the price, sending it downwards. Thus the prophesy of a peak is self-fulfilled. However, if people sell before the predicted time of the peak, it will go down early. Thus the prediction of when the peak would occur is self-refuted. Real-world trading can be more complicated than this. For instance, a firm's share price can affect its ability to raise new capital and

its ability to grow and thus justify an increased share price, a form of virtuous circle that billionaire George Soros attributes his success to (Soros,
1988, Introduction). Financial derivatives, futures and options allow
hedge funds to bet on, and try to make money from, stocks going down in
value. The presence of so many self-reinforcing and self-refuting actions,
or positive and negative feedback loops, is what makes stock markets and
economies so hard to predict.

It is because of these interdependency relations that the behaviour of
complex systems is not a sum of the behaviour of their parts. Adding one
extra person to the group may increase or decrease its total productivity.
Adding a catalyst to some chemicals may greatly alter the relative levels
of different compounds, and also alter whether they are solid, liquid or
gas. Introducing a foreign species to an island's eco-system may have a
devastating impact on other species' numbers, or have no discernable
impact at all. Similarly, introducing an innovative technology may initiate
Schumpeter's 'gales of change', rendering multiple technologies and practices obsolete while creating market opportunities for others, or it may
have little or no economic impact.

The uncertainty about such waves of changes may resemble that about
certain natural phenomena, part of the study of *self-organised criticality* (Bak, 1997). For example, earthquakes: there is no such thing as an
average earthquake. The frequency distribution for earthquakes is 'scale-
free'; it tends towards a straight line when plotted with logarithmic scales
on both axes, and mathematically takes the form of a power law: $y = A x^b$.
Power laws and scale-free distributions are among the signatures of
complex adaptive systems.

But studies of complex adaptive systems are not always about cascades
of changes. Stable patterns among the component and system states may
emerge. For example, if one combination of component states is superior
to all variations, it may come to dominate. A genome that is fitter than its
rivals, or better at replicating itself in the current environment, will tend to
increase its relative population size. A technology that is more useful and
valuable to consumers than its competing substitutes will tend to increase
market share. A solution to an organisation's problems, a strategy or
combination of actions that seems more profitable to the organisation's
members than its alternatives will tend to be adopted by the members, as
a consensus view emerges. Such emergent patterns and order from heterogeneous, adapting parts are the other well-known system-level behaviour
of complex systems.

Exactly which pattern or combination of component states emerges
may be unpredictable, the result of building on and reinforcing chance,
micro-level events. There may be several emergent patterns possible from

a particular complex system, several 'attractors' in its system state space, several 'peaks' in its 'landscape'. But that some sort of pattern or order will emerge may be predictable, as may the expected number of possible emergent patterns. Chaos theory (Gleick, 1987) showed how simple, deterministic functions could have complicated behaviour, impossible to summarise or predict without performing all the calculations. Complex adaptive systems, of simple or of complicated parts, can have relatively simple or relatively complicated behaviour – often a relatively simple change to the parts' behaviour shifts the system between the two.

Clearly, if humans live in complex adaptive systems, there are implications for decision makers and planners. Organisations are both complex systems and operate within them. The study of complexity and emergence is commonly associated with *bottom-up explanations*: that is explaining the system-level behaviour by reference to its parts' behaviour. This is not to be confused with the 'laissez-faire' approach to management, whereby leaders give their employees or citizens the freedom to act without interference from upper management, and layers of corporate hierarchies are dispensed with, in the belief that the people at the bottom will self-organise and good collective behaviour will emerge from their activities (Goldstein, Hazy and Lichtenstein, 2010, p. 4). Leaders are also components in the complex system that is the organisation; a bottom-up explanation must include them. The self-organising system includes people at all levels having to adapt to each other. If complexity science has a message for management it is more likely to be about how uncertain are the consequences of stripping away a level of organisational structure or altering the behaviour of a key component. There is no guarantee that such attempts at managerial innovation will be beneficial in their consequences. The emergent system state may be one with no organisation at all.

To sum up what is meant by complexity, there are four components: diversity, interdependencies, adaptation and emergent order. In the case of innovation, diversity means heterogeneous agents, variety of parts and stochastic variability. Interdependencies mean networks of influence, catalysts and constraints, and positive and negative feedback loops. Adaptation includes trial-and-error experimentation, organisational problem solving and social learning, natural selection, learning by doing and market pressure, all of which can lead to some changes in states being more likely than others. The existence of stable system states is determined by the presence of interdependencies between the combinatorially complex system states, a mathematical property of the complex systems that Kauffman has called 'order for free' (Kauffman, 1996). But if they are there, they can emerge, searched for by the system itself, even though none of its parts knows of them. Social–economic–technical systems collectively self-organise and search for more stable states, via relatively simple adaptive processes. Innovations emerge from the dynamics of such complex adaptive systems. This is why the study of complexity is relevant to the study of innovation. But how best to study complexity?

Research Approaches and Complex Systems

Studying innovation means studying complex adaptive social systems. Which research approaches from the social sciences will be appropriate for this? There are two familiar types of social science. The *quantitative* approach primarily involves statistical analyses of quantitative data obtained by questionnaire surveys or other numerical counting and measuring processes. A qualitative approach primarily involves constructing interpretations by reference to written and oral accounts by interviewed participants in the system and/or accounts by researchers as observers of the system, especially those observers who have participated themselves in the systems. We shall consider the quantitative approach first.

Given quantitative data one can try to propose a model, a mathematical description of some set of relations between different attributes of the items in the dataset, that *explains* the data, that is, explains in terms of some attributes why certain other attributes have the values they have. Fitting involves finding the parameters of a model that minimise the extent to which the variables to be explained deviate from what is predicted by the model's transformation of the explaining variables. The most familiar type of model is linear regression. As the name implies, this is unsuitable for systems that are nonlinear, where one component's behaviour is definitely not expected to be a linear sum of the behaviours of the other components. Regression techniques are possible with nonlinear models, but these are not widely used. There are indefinitely many mathematical functions that could be tried out, and why choose one model over another? Given enough parameters, a model can be constructed that exactly fits a dataset, but may be useless in telling us why the parameters are as they are, and unreliable in telling us how the system will behave in the future, beyond the current dataset's coverage. These are the problems of *overfitting* the data. Regression models only tell us that some variables are correlated. They do not tell us how the association has come to be. Hence, they provide no warning if the association is about to disappear (raising the philosopher's problem of induction: how can one know that a past pattern, such as the sun rising in the morning, will be repeated in the future). For that, information is needed about the *mechanisms* generating the observed pattern. As studies of chaotic functions showed earlier, the initial behaviour of some kinds of system may give little warning of its later behaviour. Small differences in initial conditions may quickly lead to large differences in outcomes – the famous butterfly effect.

In addition to these problems, many datasets cover not the entire system of interest, but only sample some of its components, or sample a dynamic system at only a few points in time. If the sampling attempts are independent of each other and concern a component whose behaviour is distributed identically each time, then statistics textbooks tell of various techniques that can be used to infer things about the system underlying the sampled data. If, on the other hand, it is plausible that the items in the sample are not independent and identically distributed, then most textbooks are less forthcoming. Also, some techniques, for example analysis of variance (ANOVA), assume that the data are being sampled from normally distributed processes. Normal distributions may apply for some phenomena that have a characteristic scale (for example, heights and IQ measures); learning the average value is of interest. But normal distributions are rarely found in empirical economic data or in innovation studies (for example, GDP, income, age and innovation generation). As mentioned already, the behaviour of some complex adaptive systems is associated with *scale-free* distributions. For these, the average value is of no use, and the variance and standard error, used for statistical tests, may be undefined.

Even when the data sources are suitable for statistical testing, researchers often draw their conclusions from tests of *statistical significance*. This is not equivalent to measures of *importance* (McCloskey, 1985; McCloskey and Ziliak, 1996). By basing decisions on statistical significance, researchers can miss important causal relations due to noisy data, while at the same time publishing claims about causal relations which if true may still be very weak. Epidemiology is another field that suffers from this problem (McCloskey and Ziliak, 2009). Given how complex the human body is, and how varied the things we do with it, one should be particularly careful in making statements about the causal effects of one lifestyle factor or about the factors behind some disease.

When studying the emergence of innovations most quantitative methods may not be appropriate anyway. Mostly they work with averages from data, but a novelty in the data is more likely to appear as an extreme case, an outlier, far from what has been typical behaviour so far. Also, when collecting data one chooses metrics and designs questions for surveys on the basis of what has been considered useful to ask in the past. Novel, emergent order, not resembling previously observed patterns, may therefore be neglected by these methods. If we have not seen it before, we might not know to ask about it.

Qualitative research methods may be able to overcome this problem. If the researcher is a part of the system, or talks to those who are, when an innovative event occurs, for example when a novel problem is encountered or a new idea is developed, the researcher may be able to trace its genesis. Human natural language is much richer than the subset employed in a quantitative survey. Even when no word exists for what is developing, metaphors and analogies may serve to construct its meaning. However, one's ability to recognise the importance of the new event depends on knowledge of other cases to compare and contrast it with. Indeed, this relatively high dependence on the subjective background of the researcher can lead many people, quantitative researchers especially, to disregard qualitative research. Also, compared to questionnaire surveys, the richer experience obtained by qualitative research can only be obtained at a slower rate and comparisons between cases take longer and are disputed more often. Extrapolating from some observed cases to a yet-to-be-unobserved case may invoke controversy. Human lives are so complex; trying to abstract and generalise from particular cases seems to threaten their complexity and autonomy. So scaling up qualitative studies is rarely performed unchallenged.

But without large-scale studies, how can we be sure that an innovation, the emergence of the next big technology, say, will be captured in detail by a researcher? With too few qualitative researchers, the chance of a researcher being in the right place at the right time seems slight. A researcher can perform a retrospective study, examining documents from the time when what we now know to have been an important technology was developed. But some documents may be missing – why keep them if you do not know how important this technology is going to be? And like war stories, tales of technological development tend to be histories of the victors.

So quantitative research, such as questionnaire surveys, can achieve large scale and the models commonly employed are easy to reapply to new cases. But the models are relatively simple and data collection is insensitive to emergence and life's complexity. Good models – models that fit the data – demonstrate correlations between variables. They do not demonstrate causal relations, or offer guidance as to what mechanisms generated the patterns, and hence provide no indication about whether the patterns of behaviour are set to continue. Qualitative research, such as analysis and critical reflection on oral and written accounts obtained from interviews and participant observation, can match the richness of social experience with the richness of natural language. But aggregating qualitative case findings and extrapolating to new ones is controversial and frequently disputed. Is there a middle way or a third option? One approach would be to try to combine qualitative and quantitative data collection. Advocates of *mixed methods* propose to do just that (Creswell and Plano Clark, 2007).

We will not explore that option in this book. Instead we will turn to the subject matter of the next section, computer simulation models.

Social simulation is the paradigm method in analytical sociology (Hedström and Bearman, 2009), whereby macro-level social facts, such as the emergence and diffusion of an innovation, are explained in terms of the mechanisms, mostly human actions and interactions, by which the social facts were generated. Pattern-oriented modelling (Grimm et al., 2005), developed in ecology, follows a similar approach. The intended contrast is with explanations of social facts in terms of their relations to other social facts, whether by statistical association or by logical deduction from grand theoretical assumptions. Analytical sociology thus bridges the gap between micro-level processes and macro-level phenomena. Economists also offered a micro-macro bridge, using mathematical integration over individuals' decisions. But it was based on homogeneous individual agents with unrealistic cognitive abilities and information, and the assumption that collectively these agents would form a system at equilibrium, so one had only to start with that state. A more responsible approach to studying complex adaptive social systems must respect our knowledge about the limits to human rationality and information sources, the diversity in human attributes and environments, and the dynamics of collective behaviour.

SIMULATING INNOVATION IN COMPLEX SYSTEMS

Why Use Computer Simulation?

The intended purpose of simulations of real-world systems is to give us something useful that we could not – for a variety of reasons – obtain from the system itself (Ahrweiler and Gilbert, 2005). An aircraft simulator, for example, gives would-be pilots an experience analogous to that of flying a plane, without the risks and costs of practising on the real thing, and with simulated, hypothetical situations that might not occur very often in real life, such as engine failures in bad weather. A key role for simulations is to answer what-if type questions, that is, simulations are not limited to the representation of the real world. Simulations of social phenomena can also save costs and avoid risks, not least the risks to professional ethics implied by experimenting on real people. Policy makers wishing to think through the consequences of their actions before they make them may appreciate being able to experience a simulation of their implementation first.

It might be wondered, however, whether computer-based representations of people can provide an adequate analogy to the real thing. Human beings can certainly be very complicated in their behaviour, and often quite mystifying. This does not stop other human beings trying to predict their actions. Indeed, humans seem to be particularly good at interpreting other humans. These attempts often fail, of course, but the failures have not been so great that we have chosen to cease the effort of making them. Part of the skill in interpreting others lies in focusing on some attributes of the person while neglecting others. Likewise, the art of modelling people requires that we leave something out. Constructing a model of a person will not make you that person, though it may lead to you performing similar actions or making similar judgments to that person and it may inspire in you the same response as you would take to the real person.

Our ability to work with mental models of people and their interactions actually goes some way to explaining how computer models of the same complex systems come to be useful. Ashby's principle that variety must match variety (Ashby, 1958) might seem to pose a problem when trying to model a complex system. To understand how simple computer models can be adequate to the task of modelling complex minds or social systems, it must be remembered that the computer models in this book are not interacting with the real-world system automatically, but only as part of modelling projects, designed, run and interpreted by human modellers. It is the combination of modeller plus model that has to meet the complexity of the real-world system (Pidd, 1999), not models in isolation, and human beings, as noted already, are particularly adapted to responding to other human beings and interpreting social situations.

Agent-based Simulation Modelling

There are various approaches to simulation, and in the next chapter we will illustrate some of them. *System dynamics* uses difference equations to represent stocks and flows (Sterman, 2000). *Discrete-event simulation* processes lists of events, with inter-event time periods determined by random sampling from particular probability distributions (Law, 2006; Robinson, 2004). In this book, however, nearly every model is an *agent-based simulation*, also known as individual-based simulation and multi-agent simulation (Axelrod, 1997a; Gilbert, 2008; Gilbert and Troitzsch, 2005).

The agent-based simulation approach explicitly represents individuals with particular attributes engaging in interactions with each other and with a shared environment. The agents are most often intended to represent people, but agents representing animals, inanimate objects, firms and countries are found within the modelling literature, and often a model will include more than one type of agent. Time steps are also explicitly represented. The attributes of the agents at one time step are determined by their attributes at previous time steps. At a time step, agent interactions occur according to some relatively simple rules of behaviour, usually represented as a few lines of computer code, for example, the rules of thumb, or heuristics, that Simon claimed were employed by human decision makers (Simon, 1955a, 1991). Agents may differ from each other in their current attributes as well as their rules of behaviour. In particular, there may be constraints about which other agents a given agent is capable of interacting with, that is, their social network. Agent behaviour may also vary according to some random elements, or stochastic processes. As a result of interactions the participants' attributes may change.

When attributes are represented visually, for example, as x and y coordinates, or as colours, a simulation user can look for on-screen visual patterns among the population of interacting agents, such as crowd formation. There will be a number of parameters to the model, controlling such features as the number of agents, the agents' initial attributes, their behaviour and their environment, and a human user may be able to learn about the model's behaviour by altering these parameters during a simulation run or between runs, using on-screen controls. Statistics can also be collected to summarise the population at a particular step of the simulation, across multiple steps during a simulation run, or across multiple runs. Such data can then be turned into charts or used in statistical tests. Thus, besides agents, agent-based modelling offers us rules of behaviour, heterogeneous agent attributes, networks, random variability, visualisation and user interaction, emergent patterns and the ability to experiment with an abstract, model system. No other simulation approach offers all this in so convenient a form. Each of the agent-based models in the following chapters will employ some or all of these features.

By comparing the output from simulation runs with varied parameter settings, the user can perform a computer simulation experiment. Given the focus of this book, we shall shorten 'computer simulation experiment' to 'experiment'. This is not an experiment on the real-world system that the model is intended to represent, but rather an exploration, made with scientific rigour, of an abstract system, albeit one which may provide some analogy to the real-world. Since experimenting with real people is often impractical, for reasons of cost, danger, ethics or lack of participants, the simulation model may be the only option we have for producing useful answers to what-if questions about the real world (Ahrweiler and Gilbert, 2005).

In other research fields and the business world, common applications for computer simulation models are fitting past data and forecasting future or hypothetical events, the most common uses for statistical models. Simulations may also be used for finding practices or quantities that produce supposedly optimal outcomes from some system, in effect a form of prediction. As will become clear, some of the social simulations described in this book have a rather different purpose. Indeed, we give demonstrations of why forecasting innovation diffusion is unlikely to succeed, because of both random variability and the complexity of the would-be adopters' contexts. Instead, our aim is often the facilitation of understanding. Often the biggest problem for the members of an organisation is not solving some problems, but rather knowing what problem they face (Rosenhead and Mingers, 2001, Introduction). Once problems have been collectively identified, structured and agreed upon, the methods for solving the problems may be straightforward. Methods for problem structuring can still be rigorous and grounded in scientific research, such as social psychology and cognitive science, and simulation modelling can be included among these methods (Robinson, 2001).

Because of this different purpose, when discussing a social simulation model, there is less emphasis on validating the model by fitting it to some historical dataset, and greater reflection on conceptual modelling, that is, the question of which concepts from the real-world system should be included in the model and which left out (Robinson, 2008a, b). Indeed, sometimes the model might not even need to be completed, that is, debugged and run, in order for participants in a modelling project to feel that their understanding of the real-world system has improved.

Compared to statisticians' models, agent-based models are likely to have many more parameters. For a statistician, this would seem to reduce their power to explain anything, since with regression models the more parameters one has when fitting data, the easier fitting becomes, and therefore the less informative it is. However, the many options and parameters in an agent-based model help make more explicit the assumptions behind the model, and allow users to focus on and experiment with the model features they think most important, rather than the features the modeller thought important at the time of programming. By including alternative functionality and optional features, the modeller aims to avoid excluding any participants in a discussion of what one can learn from the model.

Research with Simulation

The idea of using computers to simulate business processes was being written about by the early 1960s (Cyert et al., 1964; Tocher, 1963), although some of the pioneering examples of cellular automata, a form of simulated system of multiple interacting individuals, Conway's game of 'life' (Gardner, 1970) and Schelling's segregation model (Schelling, 1969, 1971) were developed initially using such non-electronic technologies as sketches in the margins of a newspaper, nickels and dimes on checkers boards and floor tiles. The growing availability of personal computers during the 1980s, and the ability to program them oneself, together with exponential rates of improvement in computer speed and storage capacity and more detailed and friendly GUIs, meant that sophisticated computer simulations could be made available to all. By the early 1990s there began to appear computer-programming languages with built-in functions developed specifically for the simulation of interacting individuals. Also during the 1990s two academic journals were launched specialising in the simulation of social phenomena, namely *Computational and Mathematical Organization Theory* in 1995 and the *Journal of Artificial Societies and Social Simulation* in 1997, since when both journals have continued to flourish (Meyer, Lorscheid and Troitzsch, 2009; Meyer, Zaggl and Carley, 2011) and more journals have been launched. Papers based on simulation modelling have also been published in mainstream journals in various fields, including sociology, psychology, environmental studies, geography, economics and business studies.

AN OUTLINE OF THE BOOK

The best-known use for models in innovation studies is that of modelling the diffusion of innovations. So this is addressed first, in *Chapter 2: The variability and variety of diffusion models*, where, following Geroski (2000), we survey several different types of diffusion model: *epidemic, probit, stock* and *evolutionary* models. Each modelling approach focuses attention on different reasons for diffusion, including: imitation of neighbours, personal responses to a changing environment and responding to the level of adoption in the population itself: a form of feedback loop. In addition, diffusion modelling provides an opportunity to compare and contrast three approaches to computer simulation: system dynamics modelling, which uses difference equations to represent stocks and flows, discrete-event simulation, which represents the occurrence of random events, and agent-based simulation, which represents individual agents, such as people, who interact with each other and an environment. Different simulation models focus one's attention on different aspect of the world one is trying to model, and can lead to different thoughts about how best to act in that world. We choose agent-based simulation over the other approaches, since it makes it easy to represent heterogeneous decision makers, as seen in the probit model, who can interact with each other according to some social network, as seen in the epidemic model, and who, potentially, could base decisions on aggregate properties, as seen in the stock model. Agent-based simulation also allows the representation of random variation, or stochastic processes, and we provide a demonstration of why it is important to include this in diffusion models. Random variability in real-world cases of diffusion, however, makes it hard to use models and past data on how many people have adopted some innovation to predict reliably how innovation adoption will behave in future.

Focusing on epidemic-type diffusion models, *Chapter 3: Diffusion and path dependence in a social network* looks at how restricting social interactions to a fixed network of relations affects the spread of innovations. Actual networks could be the reporting relations in an organisation, friendship and family ties or relations of geographical proximity. *Social network analysis* has become a familiar social-science tool in recent decades, helped by electronic means to collect data on who interacts with whom, and by the availability of PCs to calculate statistical metrics from those data. Particular interest has been paid to the structural properties of networks, including how many links there are between members of a network, how far information about an innovation has to travel between any two people, and how often a neighbour of a neighbour of oneself is also a neighbour of oneself. The first two properties affect how quickly information about an innovation travels through a networked population. The third property affects information travel whenever would-be adopters want more than one of their neighbours to have adopted before they themselves will adopt. Different networks have different structures and present these properties to varying degrees. Hence the networks have varying effects on diffusion.

Less well known is the relation between network structure and what is known as *path dependence*. This is the property of diffusion whereby adoption decisions at one time affect the chances of later adoption, particularly important when multiple innovations are spreading through the same network and competing for adopters. Our network model shows that the most likely outcome between two competing innovations varies with network structure. Another diffusion model is the *information cascades model* (Bikhchandani, Hirshleifer and Welch, 1992, 1998), in which decision makers use previous adoption decisions by others as evidence for or against their own decisions, as well as having their own, imperfect private source of information about the innovation. Under some circumstances this can lead to cascades of similar adoption decisions, when a group of decision makers start to follow an emerging consensus. In this way, the model offers an explanation of the existence of fads and fashions, and herd behaviour in crowds. But the possibility of fads undermines the utility of learning from others' adoption decisions. Therefore, decision makers should recognise when actions may be the result of herd behaviour, and discount these in their own decisions. Under such circumstances, only actions that buck the trend are likely to be counted as evidence, for they must be the result of factors other than just following the crowd. These factors could be previous decision makers following private sources of information, though they could also include mistakes and noise. Consequently, mavericks who perform surprising actions wield influence over later adoption decisions, but only while there exist at least some decision makers who are prepared to be influenced by surprising behaviour: not everyone can be a maverick all the time; a balance must be sought. This balance turns out to interact with social network structure. Applying a social network structure to a population of decision makers both reduces the threat of fads and enables them to track the current value of adopting the innovation.

Collective learning is also a feature of the models in *Chapter 4: Explore and exploit*. This begins with the view of humans in organisations proposed by Herbert Simon and colleagues in the so-called 'Carnegie School'. Humans are bounded rational decision makers, engaging in routine practices most of the time and employing rule-of-thumb search routines, known as heuristics, whenever problems call for a new combination of routine practices. James March (1991) demonstrated with an agent-based model that collectively an organisation could solve problems and learn in situations where an individual could not (Rodan, 2005). Key to this, however, was a balance in the organisation's learning practices between *exploration* of new candidate solutions, and *exploitation* of those already found. Explore too little, and you may become stuck with consensus around an inferior combination of practices, the problem of *premature convergence* (Levitt and March, 1988). Explore too much, however, and you fail to benefit from the knowledge already acquired. A later agent-based model by Lazer and Friedman (2007) builds upon the idea of this balance between exploration and exploitation, and shows that if a social network constrains who can learn from whom in the organisation, then the social network structure interacts with the search practices to determine this balance. March (1991) also points out that when in competition with other organisations, changes that raise an organisation's average or *expected performance* are sometimes inferior to changes that raise its *variability in performance* at the expense of the average. This occurs when *relative advantage* in actual performance, i.e. who came first, counts for more than *absolute advantage*, i.e. by how much they came first. This leads to a distinction between biologists and economists over what constitutes 'rational behaviour' (Slobodkin and Rapoport, 1974; Thorngate and Tavakoli, 2005), and has become increasingly important as more and more aspects of our society move towards rewarding people and businesses on a winner-take-all basis (Frank and Cook, 2010). A simple simulation of a betting game illustrates when the economists' ideal of rationality (maximising your expected utility) might not be your best course of action if your very survival is at stake (Thorngate and Tavakoli, 2005). If it involves some level of risk and cost to oneself to engage in exploring new solutions rather than exploiting existing ones, then the rewards for innovation, absolute or relative, collective or individual, become crucial.

In *Chapter 5: Science models*, some data on innovation point to these rewards being cumulative. This is one of several examples in this chapter of a relatively simple mechanism explaining a pattern observed in empirical data. Analysis of data on academic publications shows evidence of opportunities for new publication tending to go to those authors who have already published, that is a case of 'the rich getting richer'. Likewise, citations of past publications, often taken as an indicator of the quality of the cited publication, tend to reference those publications that are already rich in citations. The mark of such cumulative advantage is a *scale-free* or *power-law distribution* of the frequency of occurrence of publications per author or citations per paper. A simple simulation model can generate data that approximate such distributions, given a particular balance between processes of innovation and imitation when choosing authors or citations for new publications. Another pattern discernible in publication data is network clustering, especially clusters of authors linked by having co-authored papers together, clusters of papers linked by common keywords or topics, and clusters of papers linked by citing the same past papers. Such network clustering reflects the academic fields and subfields that give content to a publication, but is also influenced by social processes. In particular, processes that show forms of *homophily*, or the preference for similar others, can lead to clustering in networks of social interactions, especially when combined with processes of social influence, as agent-based models make clear (Axelrod, 1997b; Gilbert, 1997; Hegselmann and Krause, 2002). Science models, that is, simulations of academic production, can combine these processes to produce synthetic publication data with some of the properties observed in real data, including scale-free distributions and clustering. Could these models help inform policy concerning the organisation of real academics? The models of organisational learning in Chapter 4 were examples of simulations of knowledge creation. The processes of problem solving by heuristic search can easily be added to a science model, but calibrating it so that it still generates plausible-looking publication data is more difficult (Watts and Gilbert, 2011). An aim for such a model, however, would be to address the question of whether processes of cumulative advantage and homophily among academic authors enhances or inhibits the pace of knowledge creation.

For the science models in Chapter 5, processes were sought that would generate patterns seen in quantitative data on publications. What if we started with observations of scientists actually at work? What processes, interactions and contextual factors could we identify that might be represented in a simulation? *Chapter 6: Adopting and adapting – innovation diffusion in complex contexts* uses the literature on actor–network theory (ANT) and the social construction of technology (SCOT). Qualitative data from ethnographic studies of scientists, engineers and others at work portray innovation as a process of satisfaction of highly diverse sets of constraints. Innovative combinations of components, practices and contexts involve trade offs between material costs and physical constraints, but also the different social, political and economic interests of the parties involved. As mentioned earlier, the traditional representation of innovation taught in business schools, the so-called linear model of innovation, divides innovation into stages of first invention or the introduction of the innovation, and later diffusion of that innovation among the population. The properties of the innovation are considered fixed at its introduction. In contrast, the view of innovation to be gleaned from the ethnographic studies is that different people see different things in the innovative technology, practice or project, and they evaluate it differently. Each case of adopting an innovation involves adapting it to the new, unique context. The key to creating a successful innovation, that is an innovation that diffuses widely, seems to involve making it easy to adapt it to heterogeneous contexts. With this in mind, simulations of innovation diffusion should resemble processes of complex constraint satisfaction, rather than the simple epidemics modelled in Chapter 2. An example model is described that expresses in visual form the heterogeneity of adoption contexts. The resulting adoption patterns, however, are far from the adoption curves familiar from the literature and discussed in Chapter 2. Instead, there are interdependencies between the different components of the adoption event, and the order in which these components appear makes a difference to the outcome, that is, the simulated system shows path dependence, a concept already mentioned in Chapter 3.

Chapter 7: Technological evolution and innovation networks continues the theme of complex networks of interdependencies. Simulations of technological evolution and knowledge creation are surveyed. Various analogues are given for the creation of new technologies or knowledge, including percolation on a grid network (Silverberg and Verspagen, 2005, 2007), search for good designs of logic functions (Arthur and Polak, 2006), co-evolution, or parallel searches using genetic algorithms, of good game-playing strategies, and a search for *auto-catalytic*, or self-producing, networks of production rules, inspired by artificial chemistry and the origins of life (Padgett, Lee and Collier, 2003). The patterns generated by these models include scale-free distributions of innovation size, and the structural properties of the networks that emerge among innovation-producing firms. The models offer explanations for various empirical facts about innovation production and its relation to social organisation. Indeed, they may even lead to explanations of the emergence of new forms of social organisation itself, within which novel technologies, roles and practices are components.

The models of Chapter 7 bring together most of the principles illustrated by those described in the previous chapters. Chapter 8 concludes the book with a recap of these principles, namely stochasticity or random variability, epidemics, heterogeneous agents in changing contexts, social network structure, path-dependent outcomes, herd behaviour and the power of surprising actions, heuristic search and collective learning, cumulative advantage, homophily, innovation as constraint satisfaction, adoption as adaptation, networks of interdependent parts, co-evolution, auto-catalysis, and the emergence of innovation networks. On the basis of empirical evidence from quantitative and qualitative studies, past theorising about innovation has striven to include some of these. But rigorously incorporating them all is a task that perhaps only agent-based modelling can achieve. Employing other modelling techniques means omitting important components of the concept of innovation, and thereby weakens our power to think about innovation. It is the key aim of this book to widen awareness of the tools for thinking available to researchers working today. The global financial and economic crisis that has run during our work for this book is unlikely to be the last time innovation will have a major impact on our lives.

2. The variability and variety of diffusion models

INTRODUCTION

Once someone has invented a new product or technology or thought up a new idea, how do others come to adopt it? When someone introduces a novel practice to a group, how does it spread within that group? This chapter covers some of the modelling tools that have been developed to help us think about the diffusion of innovation – the patterns observed as ever more people in some population adopt an innovation. Two such patterns can be seen in Figure 2.1, based on data from the classic study by Ryan and Gross (1943). Shown on the left are the numbers of those who have adopted hybrid seed corn each year, from a population of 259 farmers in the mid-west United States (minus two farmers who did not adopt during the study). The numbers go up in an S-shaped curve that can be divided into phases: a lag period, during which adoption has yet to take off; a phase of rapid (near-exponential) growth; and a phase of ever slowing growth, as the population saturates with adopters. The chart on the right shows the corresponding rate of adoption – the numbers of new adopters

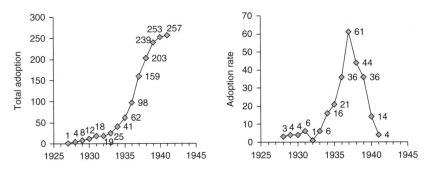

Figure 2.1   *The diffusion patterns in a classic dataset. The adoption to date (left) and adoption rate (right) of hybrid corn seed among mid-west US farmers, as published in Ryan and Gross (1943).*

in each year – with a characteristic initial rapid growth, rise to a peak and fall back to zero. Sometimes the shape of this curve may be likened to a bell, familiar from the bell-shaped normal distribution in statistics (though in this case our curve seems to be too asymmetrical and too pointy for a bell shape). These two time series – total adoption and adoption rate – form the classic diffusion patterns that a host of mathematical and computational models have tried to explain. This chapter surveys some of these, pointing out the different sources of variability within them. Three different simulation approaches are compared for their treatment of random variation in the rate of adoption. Then the behaviours of two diffusion models, the epidemic model and the probit model, are examined when would-be adopters are heterogeneous, or diverse, in their attributes. By the end of the chapter, it will be clearer why most of the rest of the book employs agent-based simulation models, and why these go beyond the functions of the most familiar model associated with innovation, the epidemic model of diffusion.

DIFFUSION AS AN EPIDEMIC

More mathematical and simulation models address diffusion than any other aspect of innovation. Of these, most take the diffusion of innovation to be like the spread of an infectious disease – an epidemic. There are several candidates for what exactly is diffusing, including the awareness that some novel technology or practice exists, the idea that it would be good to adopt that innovation now, and the knowledge essential to the application of an innovation. The second of these is the most common, appearing in marketing models such as that by Bass (1969) where – as *word-of-mouth* advertising – it is often combined with more direct forms of advertising. Without other sources of influence, the epidemic models must make some assumption concerning how the innovation originates, such as some newly arrived individual introducing an innovation from elsewhere to the population. The most basic of these epidemic models are relatively simple to understand. One can identify different phases of diffusion in their output: a period before the innovation takes off, a point at which the rate of adoption stops growing and starts to fall off, and a period when the number of adopters nears saturation of the market or population open to the innovation. If one fits the model to some empirical data on diffusion, there are model parameter values that have reasonably intuitive meanings, such as the rate of social contact between individuals in the population, and their susceptibility to this new idea or practice. Epidemic models are relatively simple to create – a spreadsheet suffices for some of them, and there are several examples on our website (see also Table 2.1 later in this chapter) – and they are also relatively easy to extend. For these reasons they represent an excellent introduction to simulation modelling.

But epidemic models are not without their problems. In this book we detail several complications to the basic epidemic models: random variation in the order of events, diversity in the attributes of the would-be adopters, social network structures constraining who may learn an innovation from whom (see Chapter 3), people's movements and changing proximity to each other (Chapter 6), and adopters' adaptations and reinterpretations of what is being adopted (Chapters 6 and 7). Epidemic models, however, are not the only approach to modelling the diffusion of innovations. Like the summaries in Stoneman (2002, Chapter 3) and Geroski (2000), we will describe some of the basic alternatives: the probit (or rank) model, the stock model, the order model and the evolutionary model, and (in Chapter 3) the model of information cascades. As Geroski (2000) points out, each modelling approach focuses attention on a different aspect of diffusion processes. In line with our contention throughout this book, these models are tools for thinking. Different models inspire different thoughts, and lead to different actions. Hence, the simulation modeller needs to be able to move beyond epidemic models, to avoid restricting the range of managerial interventions the models are used to promote.

THREE RIVAL APPROACHES TO SIMULATING EPIDEMIC DIFFUSION

In this section we do two things. First, we explain the differences between three approaches to simulating a diffusion process, for which we use a model of an epidemic. Then, we take one of the main differences between the simulation approaches – namely the use of random numbers – and show how decision makers will often need to take into consideration the effects of random variability in what mathematicians call *stochastic processes*. Approaches that model only the mean or average behaviour of a system, and neglect how that behaviour can deviate from the mean, run the risk of producing decisions and plans that are not robust to the effects of chance. All three approaches can be implemented in an Excel workbook, with charts showing their output placed alongside each other (*SI_Model_Comparison.xls* on our website).

The SI Model

Originally used to model the spread of a disease in a population from infected to susceptible people (Kermack and McKendrick, 1927), the *SI model* serves as a simple model of the spread of an idea or innovation. The mathematics of this, the logistic function, is well-known to statisticians from logistic regression and to ecologists as the logistic model of population growth. The model produces the characteristic S-shaped curve for the number of adopters to date, and a bell-shaped curve for the rate of adoption. It assumes a population of fixed size – that is, there are no births, deaths or migration during the diffusion period. Each person is in one of just two states: either *susceptible* (*S*) to infection with some innovation (novel idea, practice, or disposition to use a new technology, etc.), or infected (I), an adopter. People in the population mix freely without preference for whom they meet, and as a result of social interactions between susceptible and infected persons some susceptibles may themselves become infected, thereby increasing the availability of infected people. The model simulates this flow of people from being of type *S* to being of type *I*. Once infected, a person remains so. More sophisticated models can be built from this starting point. For example, the *SIR model* assumes that infected people can then *recover* (*R*) from the disease to become both immune to further infection and no longer infectious to others (Kermack and McKendrick, 1927). But for want of an innovation-specific analogue to such extensions we will stick to the simpler SI model.

Marketing models tend to include multiple sources of influence over people's decisions to adopt or buy a product. The Bass model, for example, represents both word-of-mouth advertising between current and would-be adopters – for which the rate and structure of social interaction between people is relevant – and the influence of advertising campaigns that provide information to everyone in the population simultaneously, continually and to an equal degree (Bass, 1969). Thus some of the eventual sales or adoptions can be attributed to the influence of the campaign, and some to word-of-mouth spreading. In Repenning's (2002) system-dynamics model there is also some managerial intervention which promotes adoption until experience of the innovation has built up sufficiently and spread via word-of-mouth to ensure it remains in the population. An SI model, on the other hand, represents the take up of an idea or innovation under the influence of word-of-mouth advertising only.

The System-dynamics Version of the SI Model

Our first simulation approach is *system dynamics* (*SD*) *modelling* (Sterman, 2000). This uses differential equations to model the changes with respect to time in variables representing some sort of stock or quantity. In the SI model we have two quantities, susceptibles (*S*) and infecteds (*I*), corresponding to the two states people can be in. The defining mathematical characteristic of the SI model is that the proportion of susceptibles becoming infected is a linear function of the current number of infecteds. This gives the following formula for the adoption rate, or number of new adopters per unit of time:

$$Adoption\text{-}rate = Constant \times (\text{\# } Infected) \times (\text{\# } Susceptibles) / (Population\ Size)$$

Population Size is assumed to be fixed and equal to the sum of the numbers of susceptibles and infecteds. In the Excel model, we simulate repeated additions of new adopters to the number to date of adopters or infected. Each addition is represented by a row of calculations. We have chosen the parameters to enable comparison with the example model in Sterman (2000, Chapter 8). The constant can be interpreted as representing a combination of how often individuals interact socially (the *contact rate*), and how probable it is that, in any one interaction between a susceptible and an infected, an infection occurs (*infectibility*). Using Sterman's example, we have a population of 100 individuals, starting with 1 infected person and 99 susceptible, a contact rate of 6 interactions per unit of time and infectibility of 0.25, giving us a constant of 6 × 0.25 = 1.5. The Excel model simulates the diffusion in steps of 4 additions per time unit. For the given parameters this results in the diffusion covering the entire population within less than 40 simulation iterations, or 10 time units. The resulting charts for total adoption and rate of adoption show smooth curves (Figure 2.2).

The Discrete-event Simulation of the SI Model

The second simulation approach is a *discrete-event simulation* (*DES*) (Law, 2006; Robinson, 2004). Unlike the SD model, which focused attention on the inter-related *stocks* of people, *S* and *I*, and the flow between them, the DES model focuses on the *event* of a person switching from

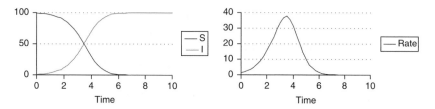

Figure 2.2   Smooth adoption (left) and adoption rate (right) curves from a system dynamics model of diffusion.

being susceptible to being infected, that is, on the adoption event. Whereas the SD model could be adding several new adopters to the I quantity with each simulation iteration, in a DES model we simulate just one new adopter per iteration. This means that for a population of 100 people, starting with 1 infected, we must simulate 99 iterations, and hence in the DES worksheet there are 99 rows of calculations. What we need to calculate in an iteration is the time that has passed since the previous event. We assume that this varies stochastically, and that at any point in time the probability of an adoption event occurring is independent of the time that has passed since the last event – that is, adoption events are 'memoryless'. This assumption, commonly used when modelling queuing systems, has the consequence that inter-event times are distributed according to a negative exponential function. If t is the time since the previous event, the probability of a particular value of t is given by:

$$P(t = x) = 1 - e^{(-x \, * \, adoption\text{-}rate)}$$

If we have a random number generator (Excel's RAND() function, in this case) producing uniformly distributed numbers in the range [0,1), we can use these as probabilities and rearrange the above formula to give us random times with the desired distribution. That is, if p is the random number, a sample time x is given by:

$$x = (-LN(1 - p) \, / \, adoption\text{-}rate)$$

The value of *adoption-rate* is given by the same formula as used in the SD model, and so depends on the constants for *infectibility* and *contact-rate*, but also on the varying numbers of susceptible and infected at that moment in time. Each time another person becomes infected, this rate will change. This, however, is the only thing that alters the distribution.

The result is that, whereas the SD model produced smooth curves, the curves for a typical run of the DES model are more uneven (Figure 2.3). This is caused by two factors. First, the SD model shows non-integer changes in adoptions – fractions of people adopting between one time step and another. The number of adoptions in the DES model can only move in integer values. Second, further smoothness is lost due to random variation in inter-event times. Sometimes one adoption follows the previous quickly; sometimes the time gap is longer. Comparing the bell-shaped curves, however, we see that, as in the SD model, the number of adoptions per time unit has a peak value when the balance between S and I quantities is 50:50. In the case of the DES model, however, exactly when this event occurs varies. This can be seen in Excel by pressing the *Recalculate* key,

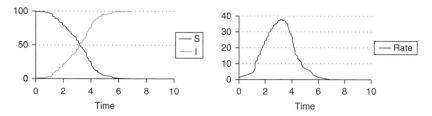

Figure 2.3   *Random variation means that the adoption (left) and adoption*
               rate (right) curves generated by the discrete-event simulation
               of diffusion are less smooth than those from a system dynamics
               model.

F9 in Windows, or *command* − = on a Mac, which forces a recalculation
of all cells, including those containing the RAND() function. With a new
stream of random numbers, the inter-event times change, and features of
the curves change, such as when the peak rate occurs and when the *S* and *I*
lines cross, and also how long it takes for the entire population to become
infected.

This means that if we want to use a DES model to tell us something
about the system being modelled, we need to run it multiple times, known
as simulation *replications*, and then work with statistics aggregated over
these times. An impression of this may be gained visually if one presses the
Recalculate key repeatedly. With each refresh the curves change shape, but
position of the peak rate varies around the point it has in the SD model.
Indeed, the SD model represents the mean behaviour for the stochastic
DES model.

The Agent-based Modelling Version of the SI Model

The third simulation approach – the one we follow the most in this
book – is an *agent-based simulation* (*ABS*). Whereas the DES model
simulated *each adoption event*, in the ABS model we simulate *each social
interaction*, not all of which need result in successful adoption. For each
interaction event we choose two agents at random, and it may be that
both are already infected or both are still only susceptible. Even if exactly
one agent of the two is infected, we still allow for the possibility – with
chance set by the *infectibility* parameter – that infection did not occur
on this occasion. The number of interactions is much larger than that of
successful infection events, and so the ABS worksheet needs a lot more
iterations – we perform 2500 rows of calculations. We could, of course,
have simulated all interaction events, including those where infection did

not occur, in a DES model. That would then have involved many more iterations, and have been comparable to the number of iterations in the ABS model. But we do not need to do this to obtain an adoption curve from a DES model, so we left them out. This option is not available when using ABS.

In both the SD and DES models, calculating the adoption rate is an essential part of the simulation. In the SD model it tells us the changes to make to the stocks. In the DES model it tells us the parameter for the current probability distribution of inter-event times. But in the ABS model it is not needed. For the sake of comparisons, we can calculate it anyway, based on the equation used before for the SD model. This represents the *expected* adoption rate at each interaction. Given an assumption that interactions occur at a fixed rate with respect to time, the resulting adoption and adoption rate curves qualitatively resemble those from the DES model (Figure 2.3), so we omit them here. As with the DES model, the SD model represents the ideal mean behaviour for the curves produced with the stochastic ABS model.

Instead, consider an alternative form of output, the actual numbers of adoptions occurring in a given period of time. For instance, for 20 time steps, we can count how many interactions have resulted in adoption since the last step. These data can be used to define an empirical adoption rate. Figure 2.4 shows example output for the ABS model, although comparable charts can be produced for the DES model. Again, hitting the *Recalculate* key will refresh the Excel workbook with new random numbers, and by doing this repeatedly the user can gain an impression of the variation in output. The empirical adoption rate tends to be much more uneven than that for the expected rate (Figure 2.3) – the result of displaying 20 points of change instead of 99. It also can produce a peak rate higher than that seen in the expected rate and in the SD model.

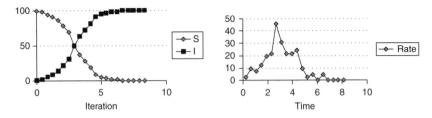

Figure 2.4   *Empirical adoption (left) and adoption rate (right) curves from an agent-based simulation. Calculating the rate at 20 points in time (instead of at every infection event) produces less smooth curves.*

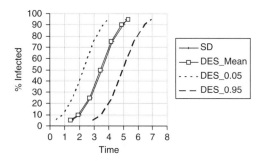

Figure 2.5   Adoption over time from a system dynamics (SD) model and from 1000 replications of a discrete-event simulation (DES), including the replications' mean and 5th and 95th percentiles. The DES can vary from the SD model to an important degree.

Why Random Variability is Desirable in Models

To summarise the three simulation approaches, SD models are deterministic: they produce the same output each time they are run, so only one run is required. The DES and ABS models are stochastic, and output statistics must be averaged over multiple runs. In addition they required more rows of calculations, with ABS the worst. Why then would we prefer either of these two to the SD model?

Stochasticity turns out to be desirable. In Figure 2.5 we show the results of running one of the stochastic models, the DES, 1000 times and aggregating its results. (The results for ABS resemble those for DES, so we omit them here.) This is the S-shaped diffusion curve, but drawn for diffusion between 5 per cent and 95 per cent of the population – in effect, from when the diffusing innovation starts to take off to when it approaches saturation point. The mean behaviour of the stochastic model is close to that of the deterministic SD model, as expected. However, we also show the 5th and 95th percentile results of the 1000 replications of the DES model. These give an indication of how much the stochastic models can vary from this mean, and suggest why one should be cautious before basing a plan on predictions from an SD model, or even from the mean behaviour of multiple runs of a stochastic model.

Suppose you only take notice of an epidemic or a new product's diffusion when infections/adoptions have reached 10 per cent of the population. How long should you expect that to take? According to the SD model it will take 2 time units. However, according to the stochastic models, there is a chance of 0.05 (or 1 in 20) that it will actually take less than 0.9 time units, and a chance of 0.05 that it will take more than 3.5. How long will it take to reach 95 per cent of the market? The SD model suggests 5.25 time units. The stochastic models suggest a 0.05 chance that it will take less than 4 time units, and a 0.05 chance that it will take more than about 7 time units.

Suppose we budget for having 50 per cent of the market within 3.5 time units. There is a 0.05 chance we will actually have less than 10 per cent of the market at this point, and a 0.05 chance we will actually have more than 90 per cent of the market. The role of stochasticity then is to play havoc with our finances!

The danger in working with mean outcomes, the so-called 'flaw of averages' (Savage and Danziger, 2012), is illustrated in Savage, Scholtes and Zweidler (2006) as follows. Consider a drunk wandering down a busy road and swaying from side to side, sometimes straying into the left lane of traffic, sometimes into the right. The drunk's *expected*, or *mean*, *position* is the middle of the road, between the lanes and where one would be safe from being hit. But the *expected state* for the drunk to be in is not safe! Wandering into one of the two lanes risks being hit, and given enough time the drunk is going to meet with an accident. This shows the difference between *a function applied to a mean* (in this case, is it safe given the mean position in the road) and *the mean of a function* (given how unsafe it is at each position taken by the drunk over time, is the drunk safe). Likewise, we would be unwise to base business plans on mean behaviour, rather than seeking plans that work well across a variety of diffusion outcomes.

Remember also that this variability came from just using the SI model, a very simplified representation of the diffusion process, with homogeneous agents and no influences other than word-of-mouth. In the real world, there is also likely to be variability in product quality and customer tastes (we assumed *infectibility* was a constant), and in social behaviour (again we made *contact rate* a constant), to say nothing of other business factors such as costs and market prices. When other factors change, a simulation model can help us react through producing new forecasts with the revised information. But forecasting those changes is itself difficult.

The Three Simulation Approaches Compared

Rahmandad and Sterman (2008) point out that incorporating a diffusion process into a larger system involving multiple factors is much quicker to achieve in an SD program than in ABS software. Indeed, Sterman (2000) contains several such models. However, when we want to model social interactions ABS has important advantages, ones that justify the extra computation we saw it needed in our Excel version.

First, we have assumed that all agents perform the same level of social interaction, and hence that they have the same chance of being selected to participate in any given interaction event. In particular, this chance does not vary with who the other participant is. Our simulated agents have no structure to their social network – everyone can interact with everyone else – and we will return to this in Chapter 3. Incorporating social network structure into an SD model is difficult – we shall not go into it in this book.

Second, once people have interacted, they each have the same chance of transmission of adoption behaviour or infection. These agents are homogeneous in their susceptibility and their persuasiveness. Incorporating heterogeneity of agent attributes – such as infectibility – in an SD model becomes harder as the number of types of agent increases. To model a population with multiple types of agent (for example, 'Technology fan', 'Technology neutral', 'Technology sceptic') we can increase the number of stocks of people and need an S and an I stock for each type. The calculation of flows between S and I stocks becomes more difficult, with more parameters to make assumptions about. (For example, does a 'Technology sceptic' who has adopted have the same persuasiveness over a 'fan' as another 'fan' does?) Even then, multiple agents are being typified by the same, representative, parameter value. In an ABS program, however, each individual agent is represented with its own attribute variables. The computation and memory storage required for this is the same, whether each agent has its own unique attribute values (heterogeneous agents) or the same values as other agents (homogeneous agents). The attribute variables can take a much greater number of distinct values. For instance, they could be continuous variables, with arbitrarily many distinct values, unlike the limited number of types in the SD model. It then become necessary to define a function for calculating infection chances based on inter-action participants' attributes, since distinct parameter values cannot now be supplied for every possible combination of attribute values. But once this is done, agents can be as diverse and unique as one wishes.

DES models fall somewhere between SD and ABS. In commercial DES packages such as *Simul8* (www.simul8.com) or *Witness* (www.lanner.com), *entities*, or agents, can be given heterogeneous attributes easily enough. But simulating a social network is more difficult. Whereas in our Excel version of DES we needed only to calculate the inter-event times for one type of infection event, a social network represents multiple interaction types, possibly as many as one for each social link or pair of agents. Given 100 agents, this is $100 \times 99/2 = 4950$ interaction processes, a prohibitive number for existing DES software.

Rahmandad and Sterman (2008) acknowledge the possibility of ABS incorporating network structures and agent heterogeneity where SD cannot, but cast doubt on the desirability of this compared with the ease with which SD models can be extended to include other factors. When it comes to the generation and diffusion of innovation, however, the impact of social network structure and heterogeneous agent attributes revealed by our simulation experiments encourages modelling at the level of individuals. Taking this in conjunction with the warning about the 'flaw of averages', we choose to concentrate on agent-based simulation modelling in the rest of this book.

ADOPTER CATEGORIES: A TALE OF TWO BELL CURVES

In the SI model we saw one way to generate an S-shaped curve for total adoption over time and with it a symmetrical, bell-shaped curve for the rate of adoption. To obtain this, the SI model included various assumptions, among them that susceptible and infected people mix without bias, and that everyone in the population is homogeneous in their susceptibility and infectiousness, or abilities to adopt and to pass on an innovation. Much innovation research, however, challenges these assumptions. In particular, when adopters have been categorised according to when they adopted, contrasts have been identified between the personal characteristics of those in different categories. In this section we use epidemic models to illustrate a conflict within three assumptions: the SI model, the concept of adopter categories and the symmetrical shape of the adoption rate curve. In fact, diffusion literature has two distinct references to 'bell curves': the adoption rate curve from the SI model and the distribution of adopter 'innovativeness', on which adopter categories are based.

Figure 2.6 shows Rogers's (1958, 2003, pp. 279–285) classic categorisation of adopters, based on the claim that 'innovativeness', the tendency to adopt innovations in general, followed a 'bell curve'. Rogers divides the bell curve up into five categories, using standard deviations from the halfway point to arrive at the following partitioning: *Innovators* (2.5 per cent), *Early Adopters* (13.5 per cent), *Early Majority* (34 per cent), *Late Majority* (34 per cent) and *Laggards* (16 per cent).

In Rogers's chapter on adopter categories he states a number of generalisations (Rogers, 2003, Chapter 7), including:

- 'Adopter distributions tend to follow an S-shaped curve over time and to approach normality (Generalisation 1).' (Rogers, 2003, p.298)

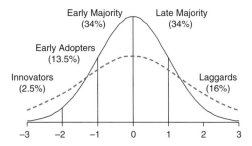

Figure 2.6   *Rogers's adopter categories, based on when they adopted.*
He assumed adoption rate formed a symmetrical bell curve –
like the normal distribution (solid line). Partitions between
categories are then sited at various standard deviations from
the midpoint. The dashed line shows the adoption rate curve
associated with the SI model, the derivative of the logistic
function. Adoption times would be more spread out with this
model.

● Relatively earlier adopters differ from later ones in socioeconomic status (Generalisations 3–7), personality variables (8–17) and communication behaviour (18–26).

Mathematically, this bell curve is distinct from that from the SI model. The sizes of Rogers's adopter categories assume the curve is a normal, or Gaussian, distribution (solid line in Figure 2.6). The adoption curve in the SI model, the derivative of the logistic function, is also shown (dashed line). While the resulting adoption rate curves may both be described as bell-shaped, the ideas and mathematics behind them are quite different. A proponent of the logistic SI model would have to modify Rogers's claim of normality.

Identifying an association between time of adoption on the one hand, and individual characteristics such as socioeconomic status, personality and communication behaviour on the other seems to promise predictive power. If they caused early adoption of a previous innovation, they may also cause early adoption of future innovations. In which case, given a population of agents heterogeneous in their characteristics, we might predict in which order they adopt from information about their characteristics.

Suppose Rogers's first generalisation is watered down to read that adopter distributions tend towards a symmetrical bell curve, rather than a normal curve. Consider again the SI model. Suppose agents' diversity in characteristics causes them to be diverse in *infectibility*, the chance

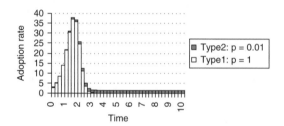

*Figure 2.7   A population with two types of agent. The first type has
            infectibility = 1, and the second type has infectibility = 0.01.
            When 50 per cent of the population are of type 1 and 50 per
            cent of type 2, the distribution is highly skewed.*

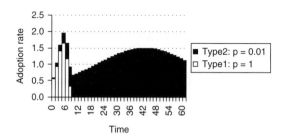

*Figure 2.8   A population with the two types of agent from Figure 2.7 now
            in different proportions: 10 per cent are of type 1 and 90 per
            cent of type 2. The result is that the adoption rate peaks twice.*

that an interaction between a given susceptible agent and an infected one
will result in the susceptible becoming infected, i.e. becoming an adopter.
Figure 2.7 and Figure 2.8 show results from system dynamics models in
which the populations consisted of agents of two types, each type of agent
having a different value for *infectibility*. An Excel workbook (*SI_Model_
WithHeterog.xls*) can be used with different choices for the proportions
of the population belonging to each type and for the *infectibility* for each
type. Depending on the choice of proportions and *infectibility*, it is possible to produce a variety of adoption rate curves. In these examples,
one adopter type has *infectibility* = 1 and the other is very different, with
infectibility = 0.01. If the two types are equal in number, this produces an
asymmetrical and upwards-skewed adoption rate curve (Figure 2.7). If 10
per cent of the population are type 1 and the other 90 per cent are type 2,
the result is a distribution of adoption over time that shows two humps, or
rises and falls twice (Figure 2.8). To obtain a truly symmetrical bell curve

from an epidemic model one would have to ensure that the population was homogeneous in its *infectibility*.

Clearly, if the distribution of real adoption times tends towards a symmetrical bell curve, as Rogers's 'generalisation 1' claims, the main process behind the diffusion cannot be that of an epidemic among agents with such a spread in infectibility. If the agents do not differ much in infectibility, then the adoption rate curve will not be so far from symmetrical. But in that case, what purpose is served by identifying the characteristics that make these agents differ in infectibility? Agent heterogeneity is intended to offer an explanation for the order in which agents adopted. Less heterogeneity means less explanatory power.

To see this, consider another simulation (*DiffusionComparisons.xls*), this time an agent-based one. As before, the population can be divided into different types, each with its own *infectibility* value. However, the diffusions of two technologies within the same population are simulated. Adoption times are output for each technology. Since an agent's infectibility is the same for each technology, there ought to be some correlation between the adoption times for the two technologies. In addition, knowing when an agent adopted one technology will provide some information about when they adopted the second. Therefore, for each technology, the program ranks the agents by their adoption times, and then classifies them for each technology as adopting either 'Earlier' or 'Later', according to whether they were among the first half of the population to adopt that technology or the among the second. If the agents are homogeneous in their infectibility, learning that an agent was in the earlier half for one technology will give one no information about where they were for the second technology. For each technology it will be 50:50 whether they are 'Earlier' or 'Later', and the conditional probability that an agent adopts 'Earlier' for a second technology, given that they adopted 'Earlier' for the first technology, will be 0.5. If adoption data shows this conditional probability to be much above 0.5, this indicates the presence of heterogeneity in *infectibility*. Figure 2.9 shows this conditional probability for various populations. In each population, 50 agents were of one type, with *infectibility* = 1. The other 50 agents were of another type, with *infectibility* set by a parameter. The x-axis shows the value of this parameter. Each data point is the mean of 100 simulation replications.

For type-2 agents' *infectibility* greater than or equal to 0.6, if you are told an agent was an 'Earlier' adopter of technology 1, your expectation that they will be 'Earlier' for technology 2 remains indistinct from 0.5, i.e. odds of 50:50. For *infectibility* = 0.1, this expectation improves to about 0.73, and for *infectibility* = 0.01, the conditional probability is over 0.92. Such predictability would be useful indeed, but recall from Figure 2.7

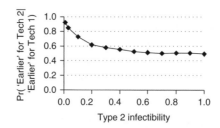

Figure 2.9   *As the infectibility of type 2 agents is decreased, it becomes*
easier to predict, from an agent being among the first 50 per
cent to adopt one technology, that they were among the first 50
per cent to adopt another technology.

above that the distribution in this case will be highly skewed, and not symmetrical like the normal curve.

Why did Rogers claim that the innovation bell curves tend towards
normality? One possibility is that studies have ended too soon: if we had
omitted some of the later columns in the histograms above, we would have
had curves much closer to symmetrical. Recall that in Ryan and Gross's
hybrid corn data (Ryan and Gross, 1943), 2 of the 259 farmers surveyed
failed to adopt. Perhaps they should be thought of as exceptionally late
adopters, so late that the data collection period failed to include their
adoption. Another possibility is that the differences in characteristics
between adopter types are significant enough to be detected in surveys, but
not large enough to distort the bell curve. But then are they large enough
to be worth studying? Do they give us predictive power with regards to
future innovation adoption?

There are, of course, other factors than differences in *infectibility* that
can affect the shape of the adoption curve. In our simulation model, altering the relative numbers of agents in each category will alter the shape –
the fewer 'Laggards' there are, the less likely they are to produce a long
tail. As we shall see elsewhere (Chapter 3), social network structure may
also impact on diffusion, while Rogers himself suggests varying marketing
strategies for each adopter category:

> The distinctive characteristics of the five adopter categories mean that these
> adopter categories can be used for audience segmentation, a strategy in which
> different communication channels and/or messages are used to reach each sub-
> audience. (Rogers, 2003, p. 299).

Perhaps by varying the messages and the initial adopters, companies
can achieve faster adoption among the 'Laggards' than would have been

expected. But how likely is it that all these factors are present in such proportions that the effect of heterogeneous adopter types is *exactly* counteracted to correct the skew and generate the symmetrical bell curve? And can the skew be counteracted without diluting away the distinctions between earlier and later adopters?

One response to this apparent conflict between the SI model, symmetrical adoption rate curves and adopters' characteristics would be to weaken Rogers's generalisation (1) still further by dropping any claim to symmetry. After all, the hybrid corn data (Figure 2.1) was not symmetrical. Another response would be to move away from the idea of diffusion as an epidemic, while retaining Rogers's generalisations. Which response one prefers in a particular case may depend on information about a system beyond that of simple adoption data. But it may be noted here that word-of-mouth advertising is not the only reason why people adopt, and the epidemic model is not the only way to produce S- and bell-shaped curves. In the next section we look at the probit model, a diffusion-of-innovations model that actually relies on adopters having heterogeneous characteristics.

HETEROGENEOUS DECISION MAKERS: THE PROBIT MODEL

We turn now to a class of diffusion models that focus not on social interactions between agents but on agents making decisions in a changing environment. Whereas in the basic epidemic model the diffusion patterns stemmed from the order in which otherwise homogeneous agents learned about the innovation, in the *probit* model, or *rank* model, the patterns emerge from a population of agents who are each aware of the innovation from the start, but may differ from each other in the adoption decision they face (Stoneman, 2002, Chapter 3). In particular, decision-making agents are heterogeneous in their perception of the costs and benefits of adopting the innovation, and their expectations and plans for the future. These latter could include known risks, the rate at which agents would discount any investment made now, and the possibility of changes to costs and benefits, such as the entry into the market by other innovative technologies. Because of these differences, some agents, faced at the same time with the same innovation in the same market, may choose to adopt while others do not.

The agents are assumed to be rational decision makers who will adopt an innovation if and only if in their view two conditions are met: that it is profitable to adopt at the current time (*the profitability condition*), and that it would not be more profitable to wait until a later time (*the arbitrage condition*). Thus we have a class of diffusion models based on common economic concepts: costs, benefits, discounting, arbitrage and rational expectations. How one values the probit model against the epidemic models comes down to whether one is willing to accept the assumptions of *rational agency*: that agents know or can calculate the utilities and probabilities of all the possible outcomes, or, if they fall short of this ideal, that use of this as an approximation to human decision making will not make any important difference to the outcomes.

Whereas the epidemic model appeals to sociologists, with its focus on social interactions, their network structure, their participants' personalities and their impact, the probit model appeals more to economists. Social network structure plays no role in the probit model, and the personalities of adopters are seldom considered; indeed, the decision-making agents modelled are more likely to be firms than individual people. Instead, data are obtained on what seem (to economists) more objectively measurable qualities of agents, including firm sizes, geographical locations and years of experience. These are then integrated with theories concerning the dynamics of costs and benefits, such as that learning how to use a technology lowers its cost, and that larger firms enjoy advantages of scale such as learning faster.

Without these dynamics, there will be no changes in adoptions, because agents only adopt when their particular conditions are met. Although a technology may have so far failed to reach the threshold price at which an agent adopts, a change in the circumstances of the technology may lead to a revised decision at a later time. Thus in the probit model, sources of adoption changes are *exogenous* to the system. In an epidemic model, agents adopt as a result of neighbouring agents having already adopted, in turn causing further neighbours to learn about the innovation. Change is *endogenous* to the system, that is, once an epidemic has first appeared in the population, no further introductions of an innovation may be necessary for the whole population to adopt it. In addition, in the simplest epidemic models adoption is a one-way transition. Having decided to adopt, agents do not then decide to abandon the innovation. But circumstances may have changed while an idea or practice is circulating: while a technology is diffusing, its costs and benefits may have altered, and knowledge of them may have improved. Such considerations are easily neglected when using the epidemic model. The probit model forces attention to be paid to the environment a technology or new idea has to survive in.

Depending on the distribution of agents' adoption threshold values, and on the dynamics of the environmental circumstances, a probit model can produce S-shaped diffusion curves and bell-shaped adoption-rate curves. The NetLogo model *ProbitModel.nlogo* on the website demonstrates some possibilities. In this agent-based simulation, a population of agents is created. Agents have two attributes. One, a binary variable, denotes whether that agent is an adopter at present. The other denotes the threshold price of the technology for the particular agent, below which that agent will adopt. Agents are initialised with threshold values sampled from a given probability distribution. We include for this distribution the options: Normal, and Log Normal. We display the agents using their threshold values as x-coordinates. This enables us to judge by eye the spread in values in the population. It also helps us to see the effect of changes in the price of the technology. Initially the technology's cost is higher than that of any agent's threshold. Over time (where time is given by the simulation iterations or 'ticks'), the cost decreases in line with a given function. We include two options for this function: 'Linear decay' (the cost is a linear function of time with a negative gradient) and 'Exponential decay' (the cost is an exponential function of time with negative exponent). If thresholds are normally distributed, and price falls linearly, the result is a symmetrical bell-shaped curve for the adoption rate. If thresholds are log-normally distributed, an exponential decline in cost will produce the same result. Davies (1979) suggested this latter scenario is at least plausible. He showed data in which firm sizes were log-normally distributed, and argued for a theory of learning in organisations whereby the cost of adoption falls exponentially over time.

As Stoneman's textbook demonstrates, probit models can be built using purely mathematical reasoning. Why then use agent-based simulation models? One contribution they can make is to introduce stochasticity: random variability in the attributes of the agents and the dynamics of the innovation's environment. In particular, the probit model assumes agents instantaneously adjust to new circumstances and the number of adopters reaches a new equilibrium before the next exogenous change. Real decision making takes time, however, and random variation in those times may undermine plans based on the expected or mean behaviour. Agents facing the same adoption decision at the same time may differ in the order in which they enact an adoption decision. This becomes important when combined with a second contribution simulation can make: interdependencies between adoption events. One firm adopting a technology can alter its market price, and hence the value of adoption to other firms, thus having the potential to alter the decision they face. So agents who differ only in the order in which they make their adoption decisions may come to different conclusions. Using an analytical probit model, rather than an agent-based version, risks neglecting the role played by adoption history. In the next section and again in Chapter 3 we review models in which small, local differences lead to the system taking very different paths.

Before that, let us reconsider the idea that agents in a probit model are like decision makers, choosing the option ('adopt at time t') that best fits their current, changing circumstances. Where do an agent's beliefs concerning costs and benefits, probabilities and the ideal discount rate come from? How do firms obtain reliable information about their circumstances? One source would be to observe what one's neighbours are doing. This reintroduces social interactions and information epidemics, and we will examine such models in the context of social networks and their structure in Chapter 3. If one's confidence about the future is partially based on the recent behaviour of one's peers and neighbours, and on claims made by them, one seems dependent once again on others, and one's fate may be determined by how well one's social group processes information about its environment. Particularly closed social network clusters may be susceptible to 'groupthink' and over- or under-confidence in their theories about the future utility of a technology. So our attempt in this section to escape from the concepts embodied in the epidemic model and the sociologists' view of the world has failed. As we noted in Chapter 1, those working in evolutionary economics and behavioural economics are prepared to acknowledge a role for heuristic decision-making methods, such as learning from others. So some sort of hybrid probit–epidemic model may be developed in the future, for which agent-based simulation may be better suited than the mathematical equation-based approaches currently employed for probit models.

ORDER MODELS, STOCK MODELS, NETWORK EXTERNALITIES AND EVOLUTION UNDER RESOURCE CONSTRAINTS

In the probit model, agents faced decisions between two competing options: adopt the innovation, or do not adopt but maintain instead your use of the old. In this section we examine diffusion models based around the outcomes of *competing technologies*. In particular we see four ways in which some technologies can gain an advantage over competitors in the level of adoption: adoption order (such as through being the first to market), scale (or stock) of adoption, network externalities (especially compatibility with other goods and services) and resource constraints. In each case, there is self-reflection: adoptions at one point in time alter the chances of adoptions later. Very often the four concepts overlap. Later chapters will exhibit these concepts in various models, so here we only introduce them briefly with the help of some case studies.

Order Models: The Importance of Early Adopters

The effects of being first to invent are well known. Copyrighting, patenting and trademarks protect the inventor against imitators who otherwise might have an advantage, since they do not have to pay for investment in R&D and have not shared the inventor's risk that an idea for a new good or service might not be technically feasible or might turn out to have no market. However, defending one's legal rights as owner costs time and money and carries the risk that the courts might not find in one's favour before one's resources are exhausted. Small firms are at a severe disadvantage here, but early sales can help by providing a revenue stream with which to meet the costs of seeing off rivals.

Other effects of being adopted early have to do with the appeal of information. Would-be adopters may be influenced by early experiences of a product, in preference to choosing later rivals for which no experience exists. Of course, the early adopters' experience may have been negative, especially if production and marketing have not been perfected. So early adoption, while influential, is not necessarily positive in its effects. In Chapter 3 we shall examine the model of information cascades (Bikhchandani et al., 1992, 1998), in which adopters employ the fact of earlier adoption in their own adoption decisions.

Sometimes it is neither invention nor adoption that a firm needs to be first at. For example, US corporation RCA was not the only company to bring radio-receiving equipment to market when it did, but it was the first to construct a working system of business functions supporting the development, production and sales of radios to US consumers (Chandler, Hikino and Von Nordenflycht, 2005).

Adoption order can lead on to the some of the other ways to gain advantage over competitors. Being first to market gives one the chance to build up scale and also increases the chance that other firms will choose your product over your rivals' when developing supporting products and services.

Stock Models: The Importance of Being Big

Stock models represent a case of cumulative advantage: those already rich in adopters get richer. In Chapter 5 we will examine this principle in models of academic science. However, rich-get-richer processes exist in business as well.

More users mean economies of scale drive the costs down, thus allowing increased profits to fund new investment and enabling the producer to
lower the price with the aim of increasing sales yet further. Revenue from
sales can also be used to fund R&D for future products, giving the firm
with the largest market share a better chance of being the first to develop
later innovations, with the order-related advantages mentioned above.
For example, RCA used its revenues from patents on radio technologies
to fund R&D into television, and the revenue from patents on black-and-
white TV technologies to fund the development of colour TV (Chandler
et al., 2005). In the 1960s the self-maintaining cycle of R&D investment
came to an end, when RCA attempted to diversify its portfolio of services
by entering the computer market to compete with IBM, and by buying up
companies whose offerings ranged from frozen food to golfing equipment,
among other things remote from consumer electronics. Perhaps as a result
of this lack of focus, Japanese companies Sony and Matsushita were able
to take over the market for TVs, and led the way in bringing video recorders to market. RCA never recovered its position, and eventually disappeared into other companies.

Reinganum's (1981) game theoretic argument, however, describes circumstances in which more users can lead to lower profit margins, and suggests that there may be an *equilibrium* number of users. Schumpeter (1943)
pointed out that when an entrepreneur takes advantage of an opportunity
to make a profit, the scale of the profit attracts others to that opportunity, and the profit margins may be competed away. Sometimes the new
entrants are firms well-established in other markets, and they bring powerful advantages of their own.

Scale can also make a firm more visible to non-rivals. Problems for a
large, household name supplier can attract the attention of journalists and
their audiences, while smaller firms go under the radar. On the other hand,
if experience of your product is positive, the increased attention that goes
with large scale can work to your advantage.

Scale also makes it more likely that other firms will choose to develop
their own products and services to fit in with yours rather than with your
rivals'. This takes us to the concept of network externalities.

Network Externalities: The Importance of Compatible Goods and Services

Scale played a complicated role in the success of RCA (Chandler et al.,
2005). Anti-trust measures by the US government forced RCA to license
its components to other manufacturers. This meant that although there
were more competitor radios in the shops, RCA would still benefit from
their sales. Indeed, it benefited from rivals exploring multiple ways of

marketing radios and TVs, without having to risk taking some of the less successful avenues itself – while the very idea of radio or TV ownership was promoted to consumers by the collective sales efforts of all these companies. So the company whose patents set the standards for radio or TV transmission enjoyed a prime position at the head of a network of technological and economic dependency relations. If rivals had not had the option of licensing RCA's technologies, whether due to government inaction, overpricing of the licenses or lack of the expertise needed to use the technologies, they might have sought alternative technological routes to developing radio and TV equipment, and perhaps found superior solutions. As it was, RCA benefitted from this network of supporting firms.

Standardisation can occur for reasons other than patenting and government intervention, due to the advantages of compatibility. Users of one product build up experience with it. Transferring this learning requires either a copy of the same product, or one sufficiently similar to, or compatible with, the first. Suppliers of supporting goods and services choose to support the product they think promises the best returns, such as the product from the industry player thought most likely to succeed.

The examples of this occurring are well known from David (1985) and Arthur (1989, 1994). Typewriters offering the QWERTY key layout prospered once touch typists emerged, and typists increasingly chose to train in QWERTY as the most popular standard, despite its having been designed originally (for already obsolete technical reasons concerning potential jams) in such a way that it slowed typists down. In the VCR market, the VHS standard pulled ahead of the Betamax standard, helped by Matsushita's greater dissemination of VHS components to other manufacturers and by the emergence of video rental stores, who started to offer more movies on VHS tapes. Computers compatible with the hardware and software standards of the IBM PC came to dominate the market for desktop business computers, even after the appearance of products with technically superior hardware, and the launch of the Apple Macintosh with its more user-friendly interface, and despite the fact that IBM's own share in the market that it had promoted shrank over time. Would-be providers of peripherals, software, training and computer know-how chose to support the standard they deemed most likely to offer returns, and as their market share grew that increasingly meant PC compatibles. A network of technologies and services emerged, each node enhancing the value of the others, with a network of provider and user firms behind it.

Thus there were *increasing returns to scale* (Arthur, 1989b). It paid to be compatible with IBM PCs, just as it did with VHS video recorders and QWERTY typewriters. That IBM failed to benefit from the scale of the

PC market owes much to the fact that the network became more important than any one of its nodes. IBM failed to incorporate in its standard-setting machines any component that could not be used or emulated by other manufacturers. Most notably it failed to take the opportunity to prevent the supplier of its operating system (OS), Microsoft, from licensing this to others. Thus began Microsoft's own virtuous circle of using the revenue from one standard OS to fund the development of the next. Even when technically superior operating systems could be said to exist – not least the free-to-install Linux systems – a desire for software compatibility maintained Microsoft's dominance.

Being first, being the largest and being in a network of supporting services are not guarantees of future success. When IBM entered the desktop PC market, the best-selling and best-supported machine was probably the Apple II, and the most-common operating system was CPM, with which Microsoft's PC DOS version 1 bore some similarity but was incompatible, for hardware reasons. So initially, software support for the new machine was limited. Thus the key factor in the case of the PC is that the market in the early 1980s was tiny relative to what it would become in just a few years' time, and many firms and consumers had yet to make a purchasing decision at all.

At this scale, complacency might be easy, such as when (according to legend) the head of CPM-supplier Digital Research chose a flying lesson over meeting with IBM executives to discuss licensing his OS software for the new IBM PC. The executives eventually chose Microsoft's product instead.

Once a few minor decisions have sufficed to push one player slightly ahead of the competition in the perception of others, the attraction of compatibility produces positive feedback loops that turn minor advantages into market dominance. In a growing market, early events turn out to wield surprising influence over later outcomes. In Chapter 3 we shall discuss several diffusion models in which the path taken by the system exhibits a dependence on early, chance events. Increasing returns to scale of adoption then cause further adoption.

Resource Constraints and Evolutionary Models

With growth based on increasing returns to scale there are constraints, including limited space in the market and scarce resources. The market cannot grow faster than the population for eventually we will all have become consumers. The firm cannot continue to grow faster than the population either, for eventually it will have recruited everyone there is. Firms tend to acquire the cheap- or easy-to-reach customers and recruits first because later ones cost more. Given the various constraints, a selective pressure operates on firms, with survival most likely to go to those best fitting the socio-economic environment. The effects can be seen in an evolutionary model of diffusion (see *EvolutionModelOfDiffusion.xls* on the website). Suppose there are two contrary practices performed within some population, such as adopting some innovative technology, or not adopting it. Suppose further that the innovation carries some advantage. Those possessing it fit the environment better, and thus run reduced risk of death, go bankrupt less often, or are less likely to feel shown up by the success of non-adopters. Whatever the scenario, the effect is for adopters to come to represent a greater proportion of the population. As adopters become the majority, however, their relative advantage is lessened; increasingly they compete with fellow adopters. Hence the rate of new adoption slows. The overall evolution in adoption described is, of course, an S-shaped curve, and in the idealised case the logistic curve already familiar from the epidemic model. The analogy with biological evolution may make one think of the more complicated pictures of evolution revealed by studies of ecosystems, and models of them as complex adaptive systems of interdependent species (May, 2001; Pimm, 2002). Likewise in the economy, firms have dependent firms (suppliers and customers) as well as their competitors, and the fortunes of any node in this web of interdependencies rise and fall in response to changes elsewhere.

The evolution and diffusion of interdependent technologies will appear in some of the models in Chapters 6 and 7, so we will not say more here. To sum up this section, models and historical cases illustrate how products benefit from being first to market, from the scale of their production and use, from previous success on the part of their owner, from standardisation, from the appeal of compatibility, from being part of a network of supporting goods and services, and from fitting their environment better than their competitors. In contrast to the probit model, where adoption events were independent of each other, and the epidemic model, where the only interdependency was social influence, we can identify many reasons for technology adoption events to affect each other. This is the domain of increasing returns to scale, positive feedback loops and market lock-in, in which relatively minor events, perhaps no more than accidents, can start self-reinforcing cycles that lead to market dominance, in spite of the relative technological merits (Arthur, 1989).

Awareness of the power of feedback loops means that one can suggest explanations for a technology's success other than saying that it must have been of superior quality to its competitors. In addition, one can predict or explain the failure of a product launched into a market in which a rival has already secured the advantages of various loops, such as becoming the market standard. But this awareness is less useful when trying to predict which, if any, of a set of new competing technologies will flourish in a new market. Often investors want forecasts, not claims that it will depend on luck.

FITTING DATA AND FORECASTING

Having now described several types of diffusion model, we turn to their potential use. Suppose we had some data on the diffusion of some innovation, perhaps the first 5 years-worth of adoption figures from Ryan and Gross's (1943) study of hybrid corn. How easy would it be to fit one of these diffusion models to the available data? If we did so, could we use it to forecast the later adoption patterns? Two considerations worth mentioning in response to these questions are the role of stochastic processes and the under-determination of models by data.

Fitting Data Output from Stochastic Processes

The Excel workbook *DiffusionFittingAndForecasting.xls* contains two epidemic models: one with stochastic processes and another without. The former will be used to generate multiple artificial datasets, on which to test our ability to fit the latter type of model. The stochastic model, a discrete-event simulation (DES), generates artificial diffusion data from given parameters. (An agent-based simulation could have been used instead, since that also contains stochastic processes. However, as argued earlier, the DES approach requires less computation, so for speed we use it now.) The parameters are taken from a model that we fitted earlier to the Ryan and Gross corn data, with a population of 257 decision makers and taking on average about 15 years to reach saturation. The second epidemic model, a system dynamics (SD) model, is then fitted to the first n years of this artificial data by adjusting its parameters. (We use the SD model for data fitting because it omits random variation, but the point below would apply to other modelling approaches.) We will try values of n from 1 to 25. For each number of data items, we fit a model with minimised mean squared error, then use the model to calculate a number of potentially interesting things to know, including: the peak adoption rate, the time (in years) when the peak adoption rate occurs, and the time when at least 95 per cent of the population have adopted. These forecasts from the system dynamics model can be compared with the 'true' values in the DES-derived dataset. We repeat this experiment 100 times, i.e. for 100 artificial datasets from the DES model.

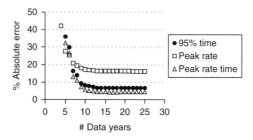

Figure 2.10   *System dynamics epidemic models were fitted to artificial adoption data generated by stochastic simulations, then various facts about the diffusion predicted. The x-axis shows how many data points were used to fit the models. The y-axis shows the absolute difference between data and model at per cent of the data values.*

Figure 2.10 shows the absolute difference between data and SD model forecast, as percentage of the data values. Clearly the more of a dataset we use to fit the SD model, the closer we get in our forecasts. However, for none of the forecast quantities do we get within 1 per cent of the data – our estimates of the peak rate differ from the artificial data by never less than 15 per cent. Before that, forecasts from just a few data points – four or five – miss the data by 40 or 50 per cent. With 10 years' worth of data points we get about as close as we can using this method. But it turns out that the average time of occurrence for the peak adoption rate is 9.4 years – so if we wait for 10 years of data then one of the most interesting aspects of the diffusion curve will have already passed! Saturation of 95 per cent takes on average 14 years, which does not leave us much time to take advantage of a forecast if we wait for 10 years' worth of data. Because of the stochastic processes involved in adoption decisions, the early time points give poor indication of what the eventual S-curve will look like. Not until adoption has taken off and the peak rate is near does the forecast become reliable, but by then it may be too late to make use of it.

The Under-determination of Models by Data

The second problem when fitting models to empirical data – under-determination – may have been realised by the reader already, as we mentioned several types of diffusion model – epidemic, probit and evolution – with contrasting theoretical underpinnings but all capable of fitting S-shaped adoption curves and bell-shaped rate curves. On the website we include six Excel spreadsheets containing attempts to fit Ryan and Gross's corn data, each file using a different theoretical model, as described in Table 2.1. The best fit achieved with each model is then compared in Table 2.2. Apart from the most basic epidemic model, which underestimates the size of the peak adoption rate, the models are rather similar in their performance. Figure 2.11 shows curves from one of them alongside that from the basic epidemic model. The data alone, then, are

Table 2.1   Various Excel-based models capable of fitting the same data.

Filename	Explanation
CornDataFit_Epidemic. xls	A basic epidemic model – the system dynamics SI model from earlier. Through random social interactions susceptibles meet adopters infected with the innovation, and become infected as well.
CornDataFit_Bass.xls	The Bass model – with parameters for a source of persuasion from outside the population ('innovation') as well as for word-of-mouth advertising within the population ('imitation').
CornDataFit_Complex Contagion.xls	What if persuading an agent to adopt requires that they have an interaction involving more than one adopter? In this epidemic model an exponent parameter represents the effect of agents having these more complex contagions.
CornDataFit_Lag.xls	The model includes a time lag between the decision to adopt and one's adoption becoming apparent to others (and hence imitable). One parameter sets the time lag.
CornDataFit_Probit.xls	Firms adopt when the cost of adoption has fallen below some threshold price. Cost is falling exponentially over time, and price relates to a firm's size. Firms are log-normally distributed in their sizes. Parameters for cost evolution, firm sizes and the firm size-price relation.
CornDataFit_Evolution. xls	A population containing two competing technologies undergoes evolution by environmental selection. One technology requires fewer resources than the other. Due to learning, this fitness advantage is increasing with the number of adopters. However, the increasing difficulty of finding improvements means that the fitness increases get smaller as adoption continues. Two parameters control fitness dynamics.

Table 2.2   Comparing the models for goodness of fit (as mean squared error), peak adoption rate, the time at which peak adoption rate occurs, and the time at which 95 per cent of the population has adopted. These statistics were collected after tuning the model parameters to minimise mean squared error.

Model	Mean Squared Error	Peak adoption rate	Time of peak rate (years)	Time of 95 per cent adoption (years)
Ryan and Gross data	N/A	61	9	12
Epidemic	56.4	39.8	9.5	14.3
Bass	17.2	56.3	9.5	13.3
Complex Contagion	14.5	56.4	9.8	12.8
Lag	22.8	60.7	9.0	12.3
Probit	21.9	55.2	9.5	13.0
Evolution	17.3	54.9	9.0	14.0

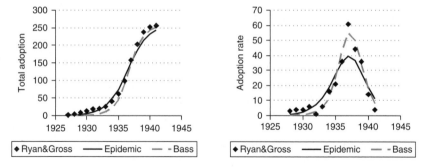

Figure 2.11   The Ryan and Gross (1943) corn data with two models fitted for adoption (left) and adoption rate (right). The epidemic model (continuous line) fails to match the peak adoption rate. The Bass (1969) model underestimates the adoption rate for the first few years. Other models tried are similarly flawed.

insufficient to decide between these models. If one has a favourite theory – based on social interactions, marketing, economic factors or whatever – then there seems to be a model to go with it that is capable of producing the S-shaped and bell-shaped curves. But the managerial interventions recommended by each theory: increasing social interactions, a media advertising campaign, reducing the price etc., are quite distinct.

THE VARIETY OF IDEAS ABOUT DIFFUSION

There have been many simulation models of the diffusion of innovations. In this chapter we have surveyed some of the major types. Different types of model focused on different aspects of diffusion.

We first investigated the importance of chance processes in diffusion, and argued for using an individual- or agent-based simulation approach, as opposed to aggregate-level system dynamics modelling. Recognising the scope for random variation in diffusion outcomes is important if robust policies are to be based on learning obtained from models. We also preferred agent-based simulation to discrete-event simulation. Even though the latter requires less computation, incorporation of social networks and other interdependencies is harder to program than in agent-based approaches.

The epidemic model focused on social interactions, and in the next chapter we will examine the role played by structure in the network of interaction patterns. Epidemic models can also include agents that are heterogeneous in their *infectibility*, or their willingness to adopt when interacting with an existing adopter. But we had problems trying to reconcile such a model with claims made by Rogers (2003, Chapter 7), based on empirical studies, that the adoption rate curve tends towards the symmetrical normal curve, and that there are differences in personal characteristics between adopters that might explain why some adopt earlier than others.

The probit model omitted social interaction, but instead raised questions of the economic environment in which adoption decisions are taken. In both epidemic and probit models, heterogeneity in the agents' attributes could make a difference to the outcome. Observations of adoption curves shaped like the normal distribution bell curve would call for explanations in terms of the kind of mechanisms that are capable of generating such patterns.

Order, stock and evolution models were also mentioned, which focus attention respectively on the order in which adoption decisions are taken, the level of adoption, and on the relative advantages of competing innovations. In each of these there are interdependencies between adoption events, positive and negative feedback loops by which earlier adoptions alter the chances of later ones. Under these circumstances, relatively minor chance events can lead to market success of one technology over another.

It was noted that when trying to decide between modelling approaches, empirical data are not guaranteed to help. Random variability means that data may deviate from the ideal, while many models are equally capable of fitting the same dataset, even when starting from different theoretical standpoints.

The diffusion of innovations is the innovation studies topic most commonly associated with modelling, and epidemic models are the most familiar diffusion model. But by now it will be clear that the topic is far from trivial. The epidemic and probit models have been easy to teach to business school and economics students, but the views of the diffusion of innovation these models encourage omit many complicating factors present in the real world. Random variability, heterogeneous agents and interdependencies can all divert system behaviour from the outcome expected following simpler, equation-based reasoning. Fortunately agent-based simulation is able to incorporate all of these sources of complexity, and models in later chapters will give examples.

In this chapter we saw how it can be difficult to find a use for empirical data, either to make useful forecasts about future adoption, or to distinguish between various competing models with diverse theoretical backgrounds. Perhaps the value of models of innovation diffusion lies not in producing accurate forecasts, but rather in recognising the diversity in people's explanations for the phenomena they see, in this case the phenomena being patterns in the adoption or non-adoption of innovations. Models allow us to state our beliefs in more rigorous, explicit and shareable form, and thus promote understanding between people, but the capabilities of a particular type of modelling can also restrict people's participation if their own mental models of the world are hard to represent in the model. Agent-based modelling, by being capable of incorporating a greater variety of assumptions about how people behave and why, is well-placed for facilitating multiple views and sources of ideas.

3. Diffusion and path dependence in a social network

WHY SOCIAL NETWORKS ARE IMPORTANT

The epidemic model of the diffusion of innovations focuses our attention on the relationship between adoption decisions and social interactions, such as happens in word-of-mouth advertising. The epidemic models based on the SI disease model in Chapter 2 all assumed that each individual in some population is able to interact freely with any other individual. Real social interactions – whether formal or informal, held in the home, in the street or at work – are not like this, however. Some social interactions are more likely to occur than others. On the day one of us types this sentence in his office in Guildford, England, there is a good chance he will bump into his co-author later and discuss the work – our offices are situated in the same building. He actually shares an office with some other colleagues, but they are not co-authors, so do not share the same interest in discussions of this book. Hence two apparent reasons for social interactions are geographical co-location or physical proximity and a preference for discussing shared interests. Geography and interests are constraining our interaction possibilities – forcing the network of interaction relations to take on a particular structure. This chapter uses simulation models of diffusion through social interactions to show that the structure of these interactions – the social network *architecture* – matters a great deal to the outcomes of innovation diffusion.

In the age of the Internet, of course, email communications are possible that overcome most geographical constraints, and we can use powerful search engines to quickly identify and take us to people online who appear to share our particular interests, without our having to pass by less familiar individuals. Nevertheless, face-to-face interactions remain important to us. From birth we are raised to engage in interaction rituals, from which come our language, culture and motivations, and our sense of self (Collins, 2004). We are also confronted daily by advertising and spend much of our time engaged with media: television, radio, newspapers, books and webpages. So these forms of communication – perhaps best thought of as corporation-to-person rather than person-to-person – also play a large part in our lives. Even on the Internet, communications still form structures – most explicitly in the case of those made via social networking sites where one has been invited to nominate with whom one is 'friends'. These online 'friendship' relations themselves form social networks, which may or may not – the question is open as we write – share the more interesting properties of networks based on face-to-face interactions. For these reasons, in an increasingly globalised world, an understanding of social networks remains essential if we are to understand the processes behind the generation, diffusion and impact of innovation.

Although studies of network structure have long been a part of the mathematics of graph theory, in the last 15 years there has been an explosion of interest in the properties of networks, driven partly by interest in the Internet, partly by the emergence of datasets based on other electronic forms of communication, and partly by computer experiments with toy models of abstract networks. In addition, increasing awareness of the importance of networks and complex systems of interdependent parts has led to networks being identified more and more in the natural world. Readable introductions to the subject include those by Buchanan (2002), Watts (2003) and Barabási (2002), though the field is advancing fast. We survey here some of the concepts of most relevance to diffusion-of-innovation models, several of which are illustrated using a NetLogo program, *NetworkDemo.nlogo*, available from the website (see appendix).

HOW NETWORK STRUCTURE MATTERS

Network Density, Degrees of Separation and the Small-world Effect

The number of links present in a network, or its density, affects both the number of nodes that can be connected, either directly or indirectly, to each other, and the ease with which information can travel between them. Figure 3.1 shows the results of constructing 100-node networks by adding links to randomly chosen pairs of nodes. One way to think about this process was described by Kauffman as the 'Buttons and Threads' model (Kauffman, 1996, pp. 54–58) and based on the model by Erdős and Rényi (1959). Suppose you start threading together randomly chosen pairs of buttons. At each point, if you picked up a button, how many other buttons would you lift up with it, via the network of threads? Each group of threaded-together buttons is a *network component*. As the ratio of threads to buttons, or links to nodes, improves, the network undergoes a phase transition from being largely unconnected – that is, each node is connected to relatively few others, either directly or indirectly – to being

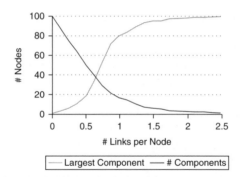

Figure 3.1   Adding links to randomly chosen pairs of nodes. As the ratio of links to nodes increases, and with it the network density also, the nodes join up to form a single component.

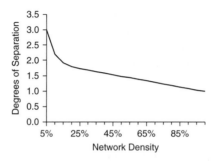

Figure 3.2   As network density continues to increase, the average degree of separation – the length of the shortest path between each pair of nodes – falls.

almost completely connected, with just one network component containing nearly all of the nodes. This connectivity has been achieved without planning – the links were added to random pairs without preference – and with a relatively small proportion of the possible links. One hundred nodes allows for $100 \times 99/2 = 4950$ possible links, or 49.5 links per node. The 2.5 links per node needed to make it more likely than not that information can pass between all 100 nodes represents a network density of 5 per cent. If we continue to add links to the network, Figure 3.2 shows that we improve the speed with which information can traverse the network, as represented by the degrees of separation, or mean length of the shortest paths between each pair of nodes, although only slowly after the network density has reached 20 per cent. Milgram's famous chain-letter experiments revealed that for some social relations the US population of hundreds of millions could be connected up with paths of average length just under 'six degrees of separation' (Milgram, 1967). This *small-world effect* is known to be present in many other types of social network, including the network of movie actors linked by their having starred in the same movie. The effect lies behind the 'Kevin Bacon game' (see www.oracleofbacon.org), whereby players try to find shortest paths in this network from a given movie actor to Kevin Bacon. So clearly a model of social network for a diffusion simulation should aim at this small-world property.

But in the real world, social relations need time and effort to be maintained. There are resource constraints on linkage. We have to balance the improvements in connectivity and degrees of separation against the cost of increased network density. The random network seen here, or *Erdős–Rényi network*, achieves its effects without any intelligence behind the addition of links. Watts and Strogatz (1998) showed that rewiring links to randomly chosen nodes could rapidly reduce the degrees of separation in a network of otherwise regular structure, such as a ring or a grid in which nodes connect to their nearest neighbours. (Figure 3.3 shows comparable results from *NetworkDemo.nlogo*.) Watts and Strogatz found that degrees of separation correlated strongly with the time required for a simulated epidemic to reach every node in the network. Random rewiring also reduces the regularity in the structure – eventually the network becomes indistinguishable from a random one – but this takes far more rewiring than is required to achieve noticeable reductions in the degrees of separation. So for an innovation to cross a network of mostly regular structure, a few long-distance

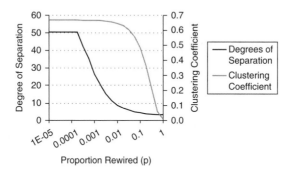

Proportion Rewired (p)

Figure 3.3   *The small-world networks of Watts and Strogatz (1998). Beginning with a network of regular structure, an increasing number of randomly chosen links are rewired. This produces first a fall in degrees of separation, then a fall in clustering. Between these two transitions, networks have both low path lengths and high clustering.*

interactions can play a large role in reducing the number of steps required. The value of longer-distance ties joining clusters had already been recognised by Granovetter (1973) when he found that information about job opportunities tended to have come not from close friends or others well connected to ones' immediate social circle, but from other social clusters via weak ties of acquaintance – the so-called *strength of weak ties.* Those in one's own social clique were likely to be strongly tied to each other and share a lot of knowledge, and so a job seeker would soon know as much as they did about current opportunities. Acquaintances in other clusters, however, connected a job seeker to more diverse sources of information.

Targeting the Hubs

The network models of Barabási and Albert (1999) identified another element capable of reducing degrees of separation – the existence of *hubs,* or nodes with particularly large numbers of links. In Chapter 1 we introduced the concept of a power-law or scale-free frequency distribution. Barabási and Albert described a method for generating networks with power-law or *scale-free* distributions of the number of links per node. (For another method for generating a scale-free distribution, see the discussion of science models and cumulative advantage in Chapter 5.) In analysing such networks, an important relation between hubs and the robustness of a network was discovered (Barabási, 2002). If randomly selected nodes are removed from such a network, the network maintains most of its connectivity for a long period. This is because most nodes selected have few links, and are fairly peripheral within the network. However, if the nodes for removal are selected with preference for those having lots of links, the network fragments relatively quickly. Scale-free networks are thus vulnerable to a removal strategy of *targeting the hubs.* The strategy of targeting the hubs reappears when considering the diffusion of innovations in the form of new information. Gladwell (2000, pp. 30–34) cites the story of Paul Revere, an individual with plenty of social connections and the personality suitable for making some more, whose ride to notify the surrounding villages that 'the Reds are coming' may have been crucial in raising support against an British army in the American War of Independence. Rogers notes that Revere also knew which individuals in each community were the best connected (Rogers, 2003, pp. 314–316). Should marketers similarly aim to persuade those with the most social connections?

Consider a network constructed using the method of Barabási and Albert, namely from an initial pair of linked nodes, a network is grown by adding one node at a time, each added node being linked to an already present node, itself chosen using sampling stratified by the number of links. Suppose the members of this social network are subject to three processes. The first is that an external advertiser targets randomly chosen members to tell them about *blue widgets*. The second process is that a rival advertiser targets nodes chosen with preference for nodes with many links, the hubs, to tell them about *green widgets*. Both types of advertiser are very persuasive, so if the targeted person has yet to adopt a widget, then they immediately adopt a widget of the advertised colour. However, widgets are also addictive. Once a person has adopted one – either blue or green – they do not change or lose their preference. The third process, called here 'infection', is that a randomly chosen node initiates a social interaction with one of its neighbours, and if the initiator has yet to adopt a widget but the neighbour has already adopted one, then the initiator imitates the neighbour's choice of widget. Using *NetworkDemo.nlogo* again, we simulate these processes by choosing at each time step, with some given chance, to simulate a process-3 interaction, and if this is not chosen, then simulating an interaction of process type either 1 or 2, chosen with equal chance. Figure 3.4(left) shows the average results of repeated experiments on a Barabási–Albert network with various values for the chance of process 3 being used (called 'Infection Chance').

With no word-of-mouth advertising in operation (*infection-chance* = 0) random targeting tends to produce more adoptions – i.e. more people adopt blue than green. But as the proportion of interactions using word-of-mouth increases, the marketing strategy of targeting the hubs becomes the more successful. The explanation is not difficult to surmise. Once a

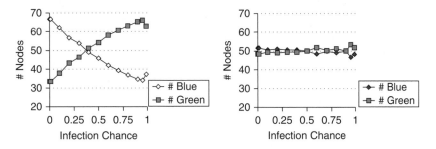

Figure 3.4　*Targeting strategies and word-mouth advertising. In a Barabási–Albert network (left), targeting random nodes ('Blue') is more effective than targeting the hubs ('Green') when no word-of-mouth diffusion is occurring, but the reverse is true when the chance of word-of-mouth infection is high. This difference is very much weaker in an* Erdős–Rényi *random network (right), though in places statistically significant at the 5 per cent level.*

hub has been targeted and has adopted, they cannot adopt again. So there is no point in targeting them again. But a strategy that targets individuals via their links is more likely than random targeting to approach the same people as before – those with lots of links. Once we introduce word-of-mouth advertising, however, the fact that hubs in the social network adopt early becomes important in two ways. First, they represent good starting points for epidemics, since they have lots of neighbours to infect and tend to be fairly central in the network. Second, their adoption blocks the passage of epidemics of rival innovations. The Barabási–Albert network accentuates this, since its tree-like structure means that once a path from one leaf node to another has been blocked by their common-hub adopting, there is no alternative path. In a network structure with lots of alternative paths, such as an *Erdős–Rényi* random network, the effect of targeting the hubs is barely detectable (Figure 3.4(right)).

Node Positions and Roles

Being a hub in the social network is not the only node role of interest to diffusion studies. Who is most able to introduce innovations to social groups? People in positions of *brokerage* have the potential to act as gatekeepers, controlling the passage of information into and out of their cluster. But their peripheral role in the network cluster, and their exposure to ideas from outside of the group, may reduce the extent to which other members of the group trust them. Advertising campaigns that target the brokers may block the word-of-mouth passage of rival ideas. On the other hand, once persuaded, the most *group-central* people may prove the best points of origin for word-of-mouth diffusion. Sociologists have debated what kind of network position provides the most social capital. For Coleman, being well-embedded in a social group leads to being trusted by its other members (Coleman, 1988, 1990). Against this, Burt (2007) argues that embeddedness promotes groupthink, and that innovations are most likely to come from brokers, who bridge the gaps in the network structure ('structural holes') between different groups or clusters. To identify these, Burt (1992) introduced the concept of structural constraint. Constraint measures the mean proportion of a node's network resources invested in particular neighbours or near-neighbours (Burt, 2007, pp. 24–27). Figure 3.5 illustrates how this varies among nodes in a modular network.

Using *NetworkDemo.nlogo* to simulate the diffusion of two rival innovations in a 100-node network we find that the effects of origin nodes' positions on the outcome vary with network structure. For each of 100 simulation runs we compared the node metric values for blue- and green-innovation-originating nodes. We also compared the final numbers of

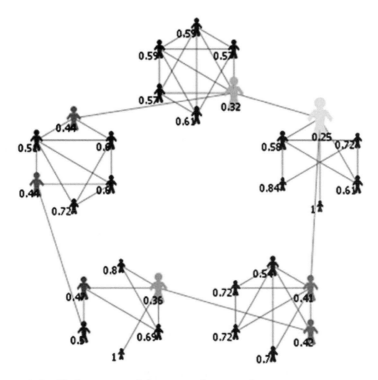

Figure 3.5   *Nodes in a modular network re-sized in inverse proportion to
their network constraint (Burt, 1992), which is also shown
in labels. The least constrained nodes act as brokers between
otherwise disconnected groups of nodes. The most constrained
have neighbours who are all neighbours of each other.*

blue and green adopters at the end of the run. Did learning that the blue-
origin has more links than the green-origin give us information about
which colour would dominate?

Table 3.1 shows the Pearson correlation coefficient between facts about
which origin had the best node metric and which innovation gained the
most adopters. (Cases were omitted wherever the two nodes were equal
in metric or in number of adopters.) The node metrics tested were: *degree
centrality* (number of links per node); *degree of separation* (average shortest path); *cliquishness* (proportion of pairs of neighbours who are neighbours of each other); *network constraint* (Burt, 1992); and *betweenness*
(proportion of shortest paths between pairs of other nodes going via that
node). There were six network *architectures*, or structures: (1) an *Erdős–
Rényi random network*, with network density of 10 per cent; (2) the *social*

Table 3.1   *Rival innovations diffuse from single origin nodes. Facts are
                collected about which origin had the better node metric value, and
                which innovation spread to the most adopters. The table shows
                the Pearson correlation coefficient between these two types of
                fact for a variety of network architectures, and node metrics.
                Numbers marked with * warranted rejecting the null hypothesis
                that the true correlation is 0, at significance level = 0.05.*

	Degree	DOS	Cliquishness	Constraint	Betweenness
Random (Erdős– Rényi)	0.377*	–0.213*	–0.122	–0.358*	0.277*
Social Circles	0.339*	–0.427*	–0.181*	–0.342*	0.258*
8-Neighbour Grid	0.7*	–0.474*	–0.7*	–0.458*	0.497*
Scale-free (Barabási– Albert)	0.072	–0.442*		–0.072	0.072
S&W 4-N Ring p = 0.1	–0.088	–0.273*	–0.078	–0.013	0.071
Scale-free (BOB 1 10 0.1 2)	0.576*	–0.479*	0.538*	–0.564*	0.408*

circles model of Hamill and Gilbert (2009, see also later), in which nodes
have random coordinates on a 100 by 100 space, and are linked to any
node within a radius of 20 (which gives a density of about 10 per cent); (3)
a regular eight-neighbour grid network in which nodes connect to those
up, down, left, right and diagonally nearest them, unless at the edge of the
grid (no wraparound); (4) the *scale-free network* of Barabási and Albert
(1999); (5) a *small-world network*, based on a four-neighbour ring with 10
per cent of the links rewired randomly, in the manner described by Watts
and Strogatz (1998); and (6) a network created by the method of Bentley,
Ormerod and Batty (2011), which approximates a scale-free distribution
of links per node while also containing clustering (unlike Barabási–Albert
networks). For more details of this last network see *NetworkDemo.nlogo*
and Bentley et al.'s paper. (The method for generating scale-free distributions in Chapter 5's science models borrows from them as well.) A brief
explanation of the parameters is that to 1 initial linked pair of nodes, new
links are added in batches of 10 a time, with chance = 0.1 that a new link
comes with a new node attached to it already, otherwise links' nodes are
chosen from those linked to in the previous two batches.

If comparison of the node metrics gives no more information about
diffusion outcomes than tossing a coin would, the correlation coefficient

will tend towards 0. If a positive metric comparison rigidly determined the outcome, it would tend towards 1. If a negative metric comparison rigidly determined the outcome, it would tend towards -1. The best values came with a grid network. Here differences in node metrics can identify whether an origin node was one of the four corners of the grid, on the edge but not a corner, or nearer the centre. Being away from the edge makes a better starting point for a diffusion epidemic. Being close to the centre is important for several other network types. In particular, degree of separation is the only indicator for the scale-free network – knowledge about which origin node was the bigger hub (i.e. had highest degree centrality or number of links) does not help. This may be perhaps because, with so few hubs to choose from, they are rarely picked as origin nodes. Hubs play an important role in reducing the mean degrees of separation in these networks, so a low degree of separation probably indicates proximity to a hub. For the Erdős–Rényi random network, however, degree of separation is of less use. The random links mean that all nodes in the network enjoy very short paths to any other node, and so differ little in their suitability for starting an epidemic. In this type of network, degree centrality and structural constraint are better predictors of diffusion success, and they work to some extent on the social circles network as well. Degree of separation was the only indicator for the rewired-ring, small-world network. All the metrics worked on the scale-free network created using the method of Bentley et al. (2011), including a positive correlation for cliquishness, in contrast to the regular eight-neighbour grid, which had a negative correlation. Low constraint leads to diffusion success in several of these networks, as does high betweenness centrality, albeit to a lesser extent. This suggests that nodes in brokerage positions are good origins for diffusion, though further investigation is required to determine whether this is because of their power to control diffusion flows, blocking the passage of rival innovations between clusters, or is because of some ability to disseminate more widely and quickly than rivals.

Complex Contagions

A less positive view of brokerage positions is implied by the work on models of *complex contagions* by Centola, Eguiluz and Macy (2007; Centola and Macy, 2007). Suppose our would-be adopters are cautious people – they will not feel confident enough of an innovation's merits unless a given number of their neighbours have already adopted it. When this threshold number of required neighbours is greater than one, the ties bridging what would otherwise be structural holes or gaps between clusters, and so-called 'long ties' connecting remote parts of social networks,

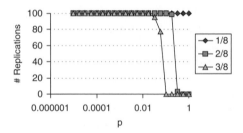

Figure 3.6   *The complex contagion experiments from Centola and Macy
(2007). Chart shows number of simulation replications, out
of 100, resulting in saturation, or diffusion to all the nodes.
If agents will not adopt until two or three of their neighbours
have adopted, the random rewiring seen in small-world
networks inhibits saturation.*

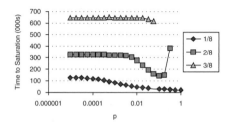

Figure 3.7   *Time in simulation iterations to reach saturation in the same
complex contagion experiments as shown in Figure 3.6. Before
random rewiring prevents saturation, and while diffusion is still
possible, rewiring can reduce the time required for saturation.*

become major obstacles. Consider a broker node with just two friends,
one in each of two social groups who themselves never socialise together –
perhaps the friends of our go-between belong to different social classes, or
live in different neighbourhoods. An innovation that has diffused widely
in one of the social groups must halt at our broker. The broker's adoption
threshold cannot be reached while only one friend has the innovation.
Even if the broker were to have sufficient friends in the adopting group,
the broker's friend in the non-adopting group would still have a threshold
of their own, and the broker's adoption alone would not suffice to cross
the neighbour's threshold. For the innovation to cross between the social
groups there needs to be a wider bridge between them – brokers with
multiple friends in both groups. Figures 3.6 and 3.7 repeat the *complex
contagion* experiments on a 10 000-node eight-neighbour regular grid in
which links have been randomly rewired to create a small-world effect (cf.

Centola and Macy, 2007, fig. 3, 5.) For each value of the rewiring chance, p.100 simulation replications were run. Figure 3.6 shows the number of these runs in which every agent had adopted by the end of the run. With thresholds of two or three neighbours, once a network has been rewired too much, the epidemic comes to a halt early. However, Figure 3.7 shows that when the rewiring does not prevent saturation, it can actually shorten the time to saturation. So for some types of diffusion process, the presence of clustering in a social network may be critical, while for other types it slows diffusion down.

The Influence of Network Structure on Diffusion

Summing up, we have seen network models in which the structure of the network, or architecture, determined whether an innovation could spread to all people, how fast it could spread and by how long a chain of adopters, what node characteristics made for a better starting point when rival innovations were diffusing, and how more complex adoption behaviour – adoption only when multiple neighbours had adopted – led to innovation failing to spread between social clusters along bridging ties. These findings came from experiments on artificially created networks, and may be sufficient to convince us that social network architecture is important to the diffusion of innovations. But this now raises the question, what structure do *real* social networks have, and which of the properties seen in the *simulated* networks pertain to real ones?

WHICH NETWORK STRUCTURE IS THE REAL ONE?

Sources of Social Network Data

We could, of course, collect some data. Network studies conducted in business organisations, for instance, send round questionnaires asking people who they interact with, who provides them with help, and even who energises or motivates them during interactions (Cross and Parker, 2004). The responses provide analysts with the relational data for drawing up maps of social networks and calculating node and network metrics like the ones we applied earlier (Wasserman and Faust, 1994). The networks contain both formal and informal communications and may be quite distinct from the organisation charts that usually define official communication channels. There are, however, problems with this approach to data collection. As with all quantitative data collection, there may be doubt over whether one is identifying the same thing each time – whether, that is, the relation claimed by A to exist between A and B is as strong and effective as that claimed by A to exist between A and C, or that claimed by B to exist between B and A. After conducting projects in over 60 organisations, social network analysts Cross and Parker (2004) advocate supplementing questionnaire data collection with interviews to check participants' understanding of the questions asked and what the results might imply. This would be particularly important when we are interested not so much in the social relations themselves but in how they might influence the generation and adoption of innovations. Further problems with self-report questionnaires include the errors introduced by people misreporting – intentionally or more often unintentionally – their social relations. Killworth and Bernard (1976, 1978) compared self-reported social links, such as are collected by questionnaire surveys, to data on actual communications gathered using unobtrusive methods, such as from radio hams and early electronic messaging systems. They found that subjects both invented connections they felt they ought to have and forgot others.

In the age of email, social networking sites and mobile phones, it might be thought easy to obtain computer records of communications. Plenty of studies of these certainly exist. But problems remain when we come to use the analysis. In the case of email, for example, we know who sent email to whom. We do not know whether the email was read, and, if so, how it was interpreted. It is very easy to copy (*cc:*) someone into an email primarily intended for another person, and send emails to whole groups of people. Consequently in the workplace we receive many emails that hold marginal or even no interest (to say nothing of spam emails) and have learnt to filter them out. For all these data sources analysts often lack information about the people who communicated. Social networking sites may have demographic data on their members, but much of it is self reported, and – as a famous *New Yorker* cartoon once put it – on the Internet no one really knows if you're a dog. Mobile phones companies have data on their customers, but it (presumably!) does not include the contents of the calls, and it certainly does not include sufficient information for us to identify innovation adoption patterns. Do people influence each other via email as much as they do via face-to-face interactions? This is an open question. Tracing the passage of an idea might be feasible in online exchanges, where messages are textual and not audio. But such content analysis is also prone to errors. If we try to automate the process, we need algorithms capable of distinguishing positive comments about some product from negative ones. We also risk missing discrepancies between what people mean by the same word or phrase. If, on the other hand, we analyse message contents ourselves, without automating the process, we cannot cover the same quantity as some of the studies of electronic communications (for example, tens of thousands of nodes). Likewise, the large quantity of data tends to discourage analysts from following up their study with alternative sources of information such as interviews with subjects. Adoption decisions may reflect interactions via rival communication channels, and exogenous influences common to members of a network, such as media advertising campaigns. If we focus on one data source we will miss these possibilities. So the Internet age has not brought any easy answers to how a social network should look or how ideas spread through it.

One type of study seems worth mentioning. In this, volunteers from some relatively small population of interest carry around electronic tracking devices that record their geographical location and motion. By comparing the movements of participants against street and building maps we get an indication of their activities during the day, while their spatial proximity to others might indicate opportunities for social interactions. It might also indicate, however, that two strangers shared a bus or elevator ride together, and made no attempt to communicate. To reduce this problem of identifying social interactions, one study conducted at Columbia University and reported to us by James Kitts (personal communication) employed devices with microphones that can detect when participants are speaking and record the durations of speaking. (For ethical reasons, they do not record the actual words spoken, and so omit contents of conversations.) Problems still remain: participants may alter their behaviour while wearing the tracking devices, and not everyone they interact with is a participant with a tracking device. So this form of data collection will not suit every kind of population of interest.

Also worth mentioning are datasets from transport networks and data on the approximate locations of people carrying mobile phones. These will not detail what face-to-face social interactions a person carried out, let alone what was discussed during them, but they will highlight how much travelling is occurring, despite it being easier than ever before to interact remotely via electronic means. That so many people travel to sit in offices together, socialise with friends or visit relatives reflects how important face-to-face interactions remain. Transport data can reveal the relative popularity of particular areas for such purposes, especially when combined with other information indicating whether the areas are largely residential, commercial or used for leisure activities.

The Attributes of Real Social Networks

Despite the problems with data collection, a few attributes of social networks can be identified from empirical studies.

Hamill and Gilbert (2009) discuss the desirable qualities of a social network model. As a starting point, they propose a model based on the idea of social circles, in which individuals connect to all within a given radius of them in some kind of 'social space'. The idea of social space is that people who know each other well are close together, and those who hardly know each other are far apart. Social space may be quite different from ordinary, physical space, although usually those who are geographically far apart are also distant from each other in social space.

As already mentioned above, social networks can exhibit the *small-world effect* – the relatively short paths between any two members of large populations – as evidenced by Milgram's chain letter experiments (Milgram, 1967). Follow-up studies by Killworth and Bernard investigated how people chose acquaintances to pass on letters to (Bernard et al., 1988; Killworth and Bernard, 1978). This means social networks are also *searchable*. Despite there being so many US citizens within six degrees of separation of them, participants were able to get letters to specific target individuals without large amounts of personal information about the targets, and without large numbers of letters. Watts, Dodds and Newman (2002) demonstrate a model for generating networks with this property of searchability, using an algorithm based on homophily, or the preference for linking to people closely related to us in interests. *SmallWorldSearch. xls* on the website is our replication of their model.

Another identifiable feature of social networks is the size of individuals' personal networks – their *degree centrality*, or numbers of contacts. It has been claimed that the network of hyperlinks between websites on the World Wide Web exhibits scale-free degree distributions (Barabási, 2002), and in Chapter 5 we will examine scale-free distributions in academic publication and citation data. But in the case of friendships and other social relations of interest, the degree distributions are neither scale-free nor follow a power law. At the one end of the distribution, very few people report having one friend or less. At the other end, rather than finding a few individuals with extremely high numbers of friends, as would be implied by a power law, we instead find evidence of an upper limit on the number of friends. Dunbar (1992, 1996) suggests this is around 150 (which has come to be known as *Dunbar's number*) and an upper limit on our brain-based cognitive capacity for distinguishing and keeping track of our friends (Hernando et al., 2010). Killworth and associates investigated the number of social relations, but placed it higher, with a mean of 291 and a probability distribution that shows a long tail at the upper end, so a few people may have much higher personal networks (Killworth et al., 2006; McCarty et al., 2001). There may be weaker social relations that permit much higher numbers of contacts, but the kinds of social relationships through which word-of-mouth diffusion might occur are probably not like them.

Social networks are *clustered*, and network analyses suggest that the distribution of clustering patterns, or *motifs* – for instance, the frequency of occurrence of different combinations of links among every set of three nodes (called *triad census*) or four nodes (*tetrad census*) – reflect the functions of the networks and the mechanisms by which they were generated. Milo et al. (2002) observed this in various scale-free networks observed in natural systems – each distinct in its pattern distribution from the networks generated by the model of Barabási and Albert (1999). Contractor's software package *Blanche* (Monge and Contractor, 2003) allowed one to link network models to theories about the processes by which links are formed. Given network data, the analysis tool *SIENA* fits advanced statistical models, including *exponential random graph models* (ERGM), based on clustering patterns (Robins et al., 2007; Snijders et al., 2006). This has been used to analyse the relation between smoking habits and friendships among teenagers (Snijders, van de Bunt and Steglich, 2010; Steglich, Snijders and Pearson, 2010), and identified the processes of 'influence' (subjects appearing to take up smoking after their friends had) and 'selection' (subjects selecting friends who shared their smoking preference). As with all statistical models, any correlations identified by the tool are not guarantees of causation, and statistical tests tell us only if the inferred parameters are likely to differ from zero, not why they do so, nor by how much. Nonetheless, this analysis tool could give useful insights when combined with theory and other data sources.

Social network analysts have long been interested in the dynamics of triads, or sets of three nodes (Wasserman and Faust, 1994, Chapter 14). Much work has been done here under the name of *balance theory*. This is founded on the idea of *structural balance* (Cartwright and Harary, 1956; Wasserman and Faust, 1994, Chapter 6) – namely that: if A has an affinity for (likes, loves, is a friend of, etc.) B, and B has an affinity for C but A dislikes C, then in so far as they know about this situation, the three individuals will feel emotionally uncomfortable about it, and changes are likely to follow, such as A trying to get to like C, or insisting to B that B break off its relationship with C. Rival accounts of why triads forms and change include those based on *propinquity* – that individuals happening to be in the same place at the same time will form social relations – and those based on *homophily* – that individuals tend to interact with those similar to themselves (McPherson, Smith-Lovin and Cook, 2001). Triad dynamics in real social networks may, of course, turn out to involve all three processes.

Whatever processes determine the formation of a network, they may continue to operate as the network evolves over time. Individuals join or leave social circles. New friendships are forged. Splits emerge between others. Better understanding of network dynamics has been requested for some time (Breiger, Carley and Pattison, 2003).

How Diffusion May Influence Networks

Changes to networks affect the generation and diffusion of innovation among those networks, but the changes may themselves be the result of innovations emerging and spreading. There are plenty of ways in which this can occur.

Policy makers and managers can influence the structure of social networks in several ways, including relocating people, instituting regular meetings and choosing who shall attend them, and installing new communications channels. Some of these initiatives may stem from ideas that diffused, such as the latest advice from business schools, or perhaps as fashions among managers who socialise with each other.

If people choose interaction partners with preference for similarity (homophily), then adoption changes the probability distribution of who will next interact with whom. As a result of diffusion, adopters of some idea become more similar to each other, while becoming less similar to those who have yet to adopt (Axelrod, 1997a, Chapter 7; 1997b).

Diffusion also means the spreading of solutions to problems. The ability of organisations to survive and the individuals within them to flourish depends in part on their problem-solving ability. In such cases, nodes that solve problems slowly or take information from bad sources are in danger of disappearing from the network. Neighbours who prove unreliable sources of information may be forgotten. On the other hand, the possessor of good information or a great new idea can attract attention, and may receive requests for interaction from new partners. Gossip about people and firms also diffuses through social networks, and reputations can influence decisions on whom to interact with.

As a result of these interactions, ideas and partial solutions may be brought together. In Chapter 7 we will return to the theme of the interplay between social networks and innovation production in the case of technological evolution. Intellectual creativity also reflects the influence of networks of social interactions. Collins's analysis of networks among philosophers and his theory of interaction ritual chains suggest that receiving attention is critical for intellectual producers to build up their emotional energy, or self confidence (Collins, 1998, 2004). Without this confidence, they hesitate to put forward new combinations of ideas. But readers and would-be disciples are limited in numbers, so in the market for ideas there is competition for this attention. Those who hesitate too much or make too quiet and modest a case, fail to attract sufficient attention to recharge themselves with emotional energy, making future successful bids for attention even less likely. Success in intellectuals' networks breeds future network position, which breeds more success. Chapter 5 will return to such processes of *cumulative advantage*.

Summing up, networks not only affect diffusion outcomes but are themselves the result of diffusion processes. Research into social networks will have to recognise the importance of their dynamics. In agent-based modelling there is a tool that can incorporate a variety of different network structures and processes of network evolution, as well as representing diffusion within those networks. We predict there are exciting simulation models to be developed along these themes. In the meantime, we hope that the reader is convinced that the structure of social networks is important for understanding the properties of diffusion of innovations, even if much remains to be discovered as to what that structure is. In the rest of this chapter we examine one final area in which network structure affects the outcomes of diffusion – namely in the production of *path-dependent* outcomes.

PATH DEPENDENCE IN SOCIAL NETWORKS

Polya's Urn Model

Consider a population in which two rival innovations are set to diffuse: green and blue technologies. At present just one person has adopted the green technology, and another person the blue, leaving all others yet to adopt but susceptible to either technology. Everyone mixes freely and is able to interact with any other. In this case, the first interaction to occur between an adopter and a susceptible is as likely to involve a green adopter as a blue one. Suppose this time it involves a green adopter, and the susceptible person is converted to the merits of green technology, before going on his or her way. Now there are two green adopters, and only one blue. The chances that the next susceptible-adopter interaction involves a green adopter, 2/3, is double that for blue. If it does indeed happen to involve green adoption, then there will be three greens to the one blue, and now a chance of 3/4 that the next adoption event involves a green. If the blue adopter now gets lucky, however, the imbalance between chances will then be slightly redressed, with three greens to two blues.

This tale of rival diffusions is known to students of stochastic processes as *Polya's urn model*. Two balls – one blue, one green – sit in an urn. One ball is drawn from the urn, its colour noted, and then it is put back in the urn, with a ball of the same colour added also. In the diffusion model version, the coloured balls stand for the adopters. Figure 3.8 shows a version of this model in Excel (*PolyaUrn.xls* on the website). 98 draws from the urn – or 98 adoption events – have been simulated, and the chart records the percentages of the green and blue balls / adopters. On this occasion the population has ended up with 71 green and 29 blue.

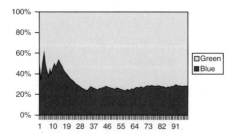

Figure 3.8   Polya's urn model as a simulation of two competing technologies diffusing through a population. Early adoption events wield more influence than later ones over the final outcome – in this case 71 adopters of green technology and 29 of blue.

With different turns of luck – that is, with another stream of 98 random numbers – we might have had a different outcome.

Early events – adoptions or draws from the urn – wield more influence over the final outcome than later ones do. The system is *path dependent*. That this is so can be seen from the jaggedness of the border between green and blue for early iterations, compared with the relatively small variations in the line from about iteration 55 onwards. The first adoption altered from 1/2 to 2/3 the chance of future adoptions of the same colour, a difference of 1/6 or approximately 0.167. The last adoption event (a blue as it happens) took the chance of more blue adoptions from 28/99 to 29/100, a difference of just 0.007. By the second half of the simulation run, the chances of a final outcome of 79 blues rather than 79 greens were now tiny, while an outcome of 99 blues was, of course, impossible.

Nevertheless, if we were to rerun our simulation many times, the probability distribution of outcomes would be uniform (Arthur, Ermoliev and Kaniovski, 1986, 1987). An outcome of 99 blues and 1 green is as likely as 50 blues and 50 greens, and 99 greens and 1 blue. This uniformity of probability distribution also holds true when we have more than two technologies involved.

This second fact turns out to be dependent on the random mixing between the blue, green and susceptible agents. Using again the *NetworkDemo.nlogo* program we can simulate the diffusion of blue and green technologies on networks of various architectures, collecting each time the final numbers of blue and green adopters. In a complete network and in an Erdős–Rényi random network the outcomes are indeed uniformly distributed. The chart in Figure 3.9a shows a histogram of results for 997 runs on 100-node Erdős–Rényi networks with density of 10 per cent (1000 replications were run, but 3 were excluded because the generated networks contained more

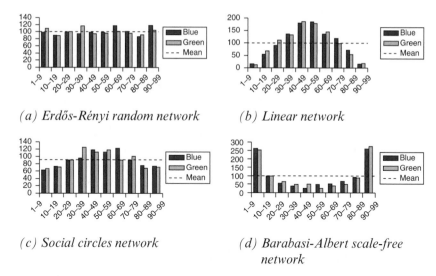

(a) Erdős-Rényi random network   *(b) Linear network*

(c) Social circles network   *(d) Barabasi-Albert scale-free network*

Figure 3.9   Frequency distributions of outcomes from 1000 diffusion simulations for four network architectures in 100-node networks.

than one connected component). The assumption that this data has come from a uniformly distributed process passes a chi-squared test of goodness of fit. Contrast that with the pattern for a linear network (b), which consists of a chain of 100 nodes, each – except for the end nodes – connected to its neighbours either side. Here there is one outcome that is more probable than all the others: 50 blues with 50 greens. Results from networks generated by the social circles method (c) of Hamill and Gilbert (2009) tend more towards the behaviour of the linear architecture than that of the Erdős–Rényi one. Results from regular grid networks were even more similar to those from linear ones, and are not shown. But the Barabási–Albert scale-free architecture (d) produces a quite opposite outcome. Here the most probable outcomes are extremes: 99 of one colour and 1 of the other. Recall that in this tree-like network, the first technology to reach a hub blocks the passage of its rival. So once again we find that network architecture makes a difference – in this case, to the distribution of diffusion outcomes.

Of course, real technologies have different properties, even when thought to be competing in the same market. Would-be adopters have varying perceptions of their value and may not adopt just the first technology that comes their way from their neighbours. In a real social network, then, we might expect the outcome to stem from more factors than just network architecture. Nonetheless, network architecture might be an important consideration when the diffusions of successions of innovations are of interest. In the next chapter we shall examine a model of organisational learning by Lazer and Friedman (2007), in which candidate solutions to some common problem diffuse through a population of agents, with each agent trying to find better variants of their current solution, either through his or her own testing of slight mutations, or through copying from one of the best solutions among his or her neighbours in some social network or organisation. Lazer and Friedman found their simulated organisations had better problem-solving performance when diffusion occurred via a linear network than via a complete one. They concluded that this was because a complete network allowed the interacting agents to converge on a single solution too fast, and thus collectively the agents failed to explore a wide range of possible solutions. (However, for another response to Lazer and Friedman's paper, see Mason and Watts (2012), who conducted online experiments using real people in artificial networks of various structures.) Once completely converged on a single solution, there is no scope for generating novel solutions through re-combination of distinct ones. Also, processes of mutation can only be applied to this single solution. Our diffusion experiments here suggest that if one wants even division of search resources between multiple rival search paths, a regular network is to be preferred to the complete and Erdős–Rényi random networks, which are in turn preferable to Barabási–Albert ones. When the diffusion of innovations really matters, as it does during collective problem solving, so too does the structure of social interactions.

FADS, HERDS AND THE POWER OF EXPERTISE

The Herd in Sync on 'the Wobbly Bridge'

The role of social networks in collective behaviour will be demonstrated in one final class of diffusion models. Before these models are described, a brief introduction to collective behaviour shown by herds and crowds is provided through the case of 'the wobbly bridge' (see http://en.wikipedia. org/wiki/Millennium_Bridge_(London) for an account, and also Strogatz, 2003, for more on synchronous action).

For the Millennium a new pedestrian bridge was constructed over the River Thames in London. Crowds of people came to walk across it on the day it opened, but the bridge soon started to wobble from side to side in what was – for the pedestrians on it – an alarming manner, and – for the architects of the bridge – something of a surprise. Later investigations suggested that one of the factors behind its side-to-side motion was herd behaviour. Human pedestrians make slight sideways pushes with their legs when walking, too slight for us to be aware of normally. But when hundreds of people simultaneously make the same sideways push in the same direction, this starts to become more important. Why would a whole crowd of individuals make pushes in the same direction simultaneously? Subconsciously at first, some individuals were able to sense slight sideways movements in the bridge's surface, which their bodies then reacted against by pushing sideways in the opposite direction. Because everyone who sensed it sensed the same bridge movement, everyone reacted in the same direction, causing the bridge to swing more noticeably, and causing yet more people to react to it. Soon, an entire bridge-full of people were walking in sync with each other, and in reaction to the swing of the bridge. This is what can happen when people's decisions are interdependent.

Dampeners were later added to Millennium Bridge and no further problems have since been reported. Investigations revealed that this was not the first pedestrian bridge to suffer from sideways swings, but news of bridges with embarrassing problems does not diffuse very readily. In addition, although civil engineers' models had long included the possibility of up-and-down movements in pedestrian bridges caused by people walking in sync, no such convention existed for modelling sideways motion – it was thought unnecessary. It would seem bridge builders are also susceptible to groupthink and herd behaviour!

Information Cascades Model and the Emergence of Fads

The diffusion models considered so far in this chapter are of the epidemic type – agents adopt because their neighbours have adopted. We saw in the previous section how this interdependency between adoption decisions led to path dependent outcomes. What if agents have some external source of information about an innovation, as well as being aware of previous adoption decisions – their own decisions or those by others – and they incorporate both sources of information into their current decision making? The models of information cascades (Banerjee, 1992; Bikhchandani et al., 1992, 1998) explore how trying to learn from others can lead to herd behaviour and fashions and fads – that is, rises and falls in the popularity of an innovation that have little or no connection to its actual value. In this section we introduce the ideas behind the model, then extend them in the next section by considering what happens when the others whose decisions one learns from are neighbours in a social network.

The Excel workbook *Cascade.xls* covers several scenarios described by Bikhchandani et al. (1992, 1998). Suppose there is a population of agents, each of whom will have to make a binary decision, such as whether to adopt some innovation or not. The agents make their decisions in turn

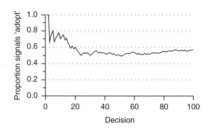

Figure 3.10	*The cumulative proportion of private signals recommending adoption. As more and more signals are considered, the proportion converges on the true bias towards adoption 'adopt', namely 0.51.*

in some arbitrary order, and each agent has information on all previous agent decisions. In addition to this information, each agent has a private source of information about the innovation itself, accessible only to that agent, but this comes via a potentially unreliable method, 'a very weak signal', which we might think of as representing some experiment or observation, or communication from someone outside the system. Let us model this signal as showing a slight bias towards adopting, by modelling it as a biased coin toss (i.e. a Bernoulli-distributed random variable) with probability 0.51 of coming up 'heads' or 'adopt'. If each agent were given information on previous agents' signals, then the proportion of these signals set to 'adopt' would be his or her best estimate for the chance that 'adopt' is indeed the best, and the worksheet 'ObservingSignals' shows that this proportion tends to something a little over 0.5 quite quickly (Figure 3.10).

Now suppose agents do not have information on previous *signals*, but instead they know previous agents' *decisions*. The first agent – following the naming from Bikhchandani et al. (1998), we will call him Aaron – has information only on his signal. So he uses that alone – if it says 'adopt', he adopts; if not, not. The second agent, Barbara, has her own signal, plus the outcome of Aaron's decision. We shall assume that all agents believe everyone's signals to be equally reliable. (We leave it to our readers to decide if that assumption holds true for any organisation they know!) If Barbara's signal matches Aaron's decision, then all her information sources point to the same action and she follows that. If, however, the sources conflict, then Barbara can decide between the options by some extra, unbiased method, such as tossing a coin. The third agent, Clarence, has information on his signal, and on Aaron and Barbara's actions. Again, if his signal matches the previous actions, then he chooses that same action. But if Aaron and Barbara agreed, then the evidence of their two actions outweighs that of

Clarence's one signal, so Clarence must follow the preceding consensus, even if his signal contradicts it. Only if the difference between preceding numbers of 'adopt' and 'do not adopt' actions is less than two does Clarence have a use for extra information. If numbers of 'adopt' and 'do not adopt' actions balance each other, Clarence uses his signal to decide. If the numbers differ by one, then if Clarence's signal agrees with the slightly more popular action, he follows that, and if his signal disagrees, then following it would create an equal balance, so in this case he uses a coin toss to decide. Later agents follow the same decision rules as Clarence, but with one important difference. Like him they can see the decisions by Aaron and Barbara that his decision was a response to. If Aaron and Barbara agreed with each other, then Clarence – if he is rational – must have followed their decision, irrespective of what his private signal was. But in that case his decision gives later agents no new information concerning whether or not to adopt. So later agents base their decisions on the evidence of Aaron and Barbara's actions alone, which override private signals, and thus the later agents' decisions also make no contribution to the evidence base. A cascade has begun of agents who blindly follow the action from the first two agents – each new agent follows the herd. Even if Aaron and Barbara's decisions were to contradict each other, a cascade typically begins after just a few more agents' decisions – just as soon as the absolute difference between numbers of 'adopters' and 'non-adopters' reaches two agents. Refreshing the random numbers in the worksheet 'ObservingActions' (press *F9* to do this) will show that these cascades start very quickly – usually after two or three agents' decisions. In addition, the direction taken by these cascades shows little preference between 'adopt' and 'do not adopt' for the state the herd follows. The fact of a cascade seems no indication of the value of adopting against not adopting the innovation (Figure 3.11). It

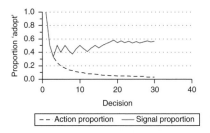

*Figure 3.11   Comparing the proportion of private signals with the
proportion of actions taken. Early 'do not adopt' signals
have started a cascade of 'do not adopt' decisions, and later
signals, which bring the trend back towards the true value of
0.51 have no influence over decisions.*

demonstrates the importance of the order in which signals arrive: the early agents' signals are critical; the signals for later agents are not consulted and have no influence.

Mavericks and the Power of Surprise

These cascades seem too extreme to form good representations of real herd behaviour – after all, in reality not everyone follows a fad. Suppose, therefore, there is some given chance that an agent decides to ignore all information about past actions and rely on his or her signal alone. Perhaps the agent is a new arrival to the area or organisation, or an expert who feels no need to learn from others, or just a maverick with high self-confidence. On the 'ObservingActions' worksheet we can simulate this possibility by entering a non-zero chance of using one's own signal ('Method2') instead of the 'Learning' from others ('Method1'). Whenever the signal is used, if it contradicts the herd opinion, then the cascade is halted for one agent turn. Now this may come as a surprise to other agents – if they were assuming that the agent was like the others, they would be expecting him or her to follow the herd. By contradicting the herd, the agent acts as if he or she is basing a decision on information other than just the current fashion. Thus the surprising action may be evidence of a private signal, and later agents can add it to the running totals, the evidence on which they base their decisions. Usually thereafter, the cascade recommences within a turn or two – the evidence from the surprising result has reduced the difference between 'adopt' and 'do not adopt' counts to one, but the next agent's private signal is as likely as not to take us back over the threshold to a difference of two. To produce a reverse cascade would require that a sequence of agents used their signals and that their signals tended to counter the fad. Figure 3.12 shows one such case. There have been surprising decisions at four points. Three produce no more than a kink in the action history trend, but the fourth has led to a reverse trend heading up. The chance of a cascade or reverse cascade heading in the right direction improves when the signal chance is much clearer – for example 0.8 instead of 0.51. Then if a cascade has headed towards 'do not adopt', a surprise interruption allows extra evidence to enter, and each extra signal is likely (with chance 0.8) to be towards 'adopting'.

Bikhchandani et al. (1998) suggest this potential to halt a fad makes outside experts and mavericks a valuable addition to social groups, and rational agents who learn from others should therefore seek to identify such individuals and weight the evidence provided by their actions. But how easy is it to identify these individuals from their actions? Suppose the information channel that tells agents about preceding actions is not

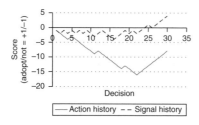

Figure 3.12   *The cumulative histories of adoption actions and private*
signals, where 'adopt' scores +1 and 'do not adopt' score -1.
The action history shows a cascade of 'do not adopt' actions.
At three points surprising actions produced brief kinks in this
before surprise at a fourth point produces a reverse cascade.

100 per cent reliable – that is, that there exist some non-zero chance that
information on a particular action will be reversed. Depending on the
chance of this 'noise', this effect can slow a cascade or even prevent it
from forming altogether because if agents' decisions do follow a fad, they
are then reversed, thus reducing the apparent popularity of the faddish
behaviour. So, arbitrary decisions can wield power over the herd just as
independent experts' decisions could – the key element is that of being
surprising. If the chance of noise is too strong, the history of actions taken
is distinct from that of using independent experts. Set the chance of relying
on one's signal alone to 1, and the proportion of decisions resulting in
'adopt' becomes identical to the proportion of signals resulting in 'adopt'.
Set instead the noise chance to 1, and the 'adopt' decisions proportion
tends to 0.5.

Away from these extreme cases, however, the similarity between independent expertise and arbitrary decision undermines the power of the
mavericks. They could be experts acting in accordance with independent
sources of information to which we lack direct access, or they could be
wildcard characters behaving in an arbitrary manner, or even be bloody-
minded contrarians, determined to oppose whatever the majority have
been doing. With this in mind, how much influence should we allow
those who act surprisingly? This question may be resolved by attention
to context. Do we have other information about the maverick that might
support one explanation of their behaviour over another – for example,
that they possess a university degree in a relevant subject, or 10 years
of experience in this field? However, organisational studies suggest that
if one's background knowledge is too clear to see, one actually *loses*
power. In particular, there appears to be a connection between areas of
uncertainty and the wielding of power (Collins, 1992, p. 83; Wilensky,

1964). Taking Wilensky's example, consider a mechanic and a doctor. If a mechanic fails to mend your car by a certain date, you find another mechanic the next time it has a fault. But if a doctor fails to cure you from one appointment, you blame the disease, not the doctor. Individuals whose expertise is in a simple area wield little power. The more predictable their decisions, the easier it will be for their clients or managers to internalise the decision-making process themselves, and thus be able to bypass the expert. By contrast, professionals in *uncertain* areas, including health, war, the economy and many business areas (and simulation modelling projects?), are able to maintain considerable status. We can neither afford to ignore their recommendations and decisions, nor blame them if the first few resulting actions turn out to be bad.

In conclusion then, deciding whether or not to adopt an innovation, based on the information provided by past adoption decisions, is problematic. The simulation model of information cascades suggests that learning from others leads to the formation of fads. Agents who give equal weight to all decisions will be susceptible to faddish behaviour and follow the herd, even if they are aware of the possibility of herd behaviour and discount the actions of others that seem in accordance with the fad. Agents who produce surprising decisions can draw attention to themselves, and wield some power in pausing or even reversing the herd. But distinguishing their surprising decisions as new evidence rather than arbitrary noise is hard. It requires much more information than that provided by the isolated adoption decisions in our model. Evidence from organisational studies suggests we give greatest influence to those with unique access to expertise in some area of uncertainty.

SOCIAL NETWORKS AND HERD BEHAVIOUR

With the discussion of access to expertise in the last section, one might start to think about social networks again – recalling the role played by brokers who bridge the structural holes between clusters. What if one adds social-network considerations to the information cascades model? In this section we show that once again social network structure makes a difference to adoption diffusion.

A population of agents is connected via some social network of fixed structure. Each time step, one randomly chosen agent makes a decision whether or not to adopt some innovation. Each agent maintains a personal record of the evidence known to it, for and against adoption. This is updated as follows:

An agent records a decision outcome if and only if:

- *either* the decision was taken by that agent on the basis of its private signal;
- *or* the decision was taken by one of the agent's neighbours, and seemed *predictable* on the basis of the agent's memory (i.e. the absolute difference in evidence of 'adopt' and 'do not adopt' was less than two), but the actual decision outcome was then *surprising* to the agent.

As before, our agents will occasionally, by chance, skip this evidence and jump straight to consulting their private signal only, rather than learning from others.

In the original information cascades model, every agent knows about every previous decision made by any other agents. The analogue to this in our new model – *NetworkInfoCascades.nlogo* – is to arrange our agents in a complete network. Figure 3.13a shows typical results, with the chance of learning from others set to 0.9, and a bias from the private signals towards adoption of 0.6. Fads emerge, with adoption (represented by 'blue-infected') coming in and out of fashion in opposition to non-adoption ('green-infected'). If we instead arrange the agents in an Erdős–Rényi random network with density of 0.1, the fads disappear (Figure 3.13b). Instead, a consensus emerges quickly and is maintained thereafter – in this case, in favour of adoption, as might be expected given that the private signals were biased so. In fact in repeated experiments, random networks were always observed to produce a consensus position, though in a small minority of cases the consensus was in favour of the 'wrong' action – i.e. against the private signal bias. Agents given other network architectures were likewise prone to picking one consensus and sticking with it, though social circles and 8-neighbour grids did not show evidence of the population ever preferring the wrong action.

Which might be the best network architecture for agents employing a mixture of private signals and information about neighbours' decisions? To estimate this, we modified the private signal process so that every 20 000 time steps (or decisions) the bias was reversed: from 0.6 to 0.4; then back again. Figure 3.14 shows two such runs on different network structures. Often the Erdős–Rényi random network is seen to be slower than a regular grid in responding to a change in signal bias.

Each time an agent made a decision contrary to the current bias, this was counted as a 'bad' decision. At the end of a simulation run of 100 000 decisions, involving five switches of bias, the per cent of decisions that were bad is calculated. Figure 3.15 compares four architectures across a range of values for the learning-chance parameter: complete network; Barabási–Albert scale-free; Erdős–Rényi random; and an 8-neighbour regular grid.

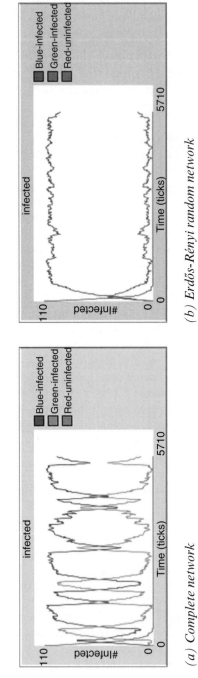

(a) Complete network

(b) Erdős-Rényi random network

Figure 3.13   Charts show adoption (blue) against non-adoption (green) in a 100-node network with the chance of learning from others set to 0.9. The information cascades model shows fads when agents learn from others in a complete network (a), but a consensus is maintained when agents learn only from neighbours in a random network (b).

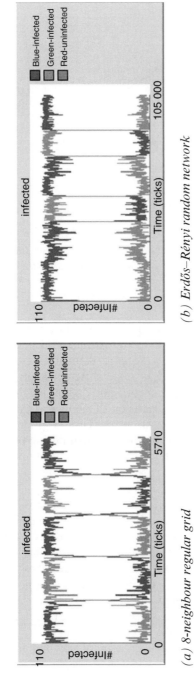

(a) 8-neighbour regular grid

(b) Erdős–Rényi random network

Figure 3.14 If the information bias reverses every 20 000 ticks, can the population track this with their adoption
decisions? Consensus switches within a couple of thousand ticks when agents are in an 8-neighbour regular
grid (a), but can sometimes take much longer on an Erdős–Rényi random network (b).

95

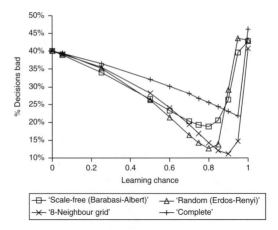

Figure 3.15   *How often should we learn from others when making adoption*
                 decisions in a social network? x-axis shows the chance of
                 agents learning from others rather than using private signals
                 about whether or not to adopt. y-axis shows what percentage
                 of their decisions go against the current bias in private signal.

Also tested were the social circles network and linear network, but these
gave performances quite similar to the regular grid and the Barabási–
Albert networks respectively, so we omit them from the charts for the
sake of clarity. What we see is that we get different levels of performance
depending both on network architecture and on how much use we make of
learning from others. To have at least some tendency to ignore past actions
leads to much better tracking of the current best adoption behaviour. But
the ideal amount of learning from others is still quite high. What that ideal
is varies between network architectures. The complete network – which
produced the fads – is by far the worst performer. The regular grid, Erdős–
Rényi and social circles networks were among the best, but they may differ
in their ideal chance of learning from others, with Erdős–Rényi networks
doing best at a lower chance of learning (0.8) to the other two (0.9). Why
the differences? The complete network may be suffering from too many
agents sharing the same observations of past decisions. Experiments on
Erdős–Rényi random networks with varying densities would test this
hypothesis. Between the other networks, a key difference is clustering –
or the extent to which my neighbours are neighbours of each other. This
is very low for Erdős–Rényi architecture, zero for Barabási–Albert, but
quite high for many of the nodes in networks organised in line with some
kind of spatial regularity, as social circles networks and grids are.

We repeated our experiment with an extra factor – that of 'noise'

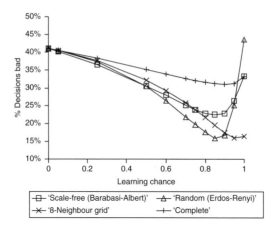

Figure 3.16   We repeat the experiment from Figure 3.15 with a chance of 0.05 of 'noise' causing agent decisions to be reversed when recorded.

(Figure 3.16). With a chance of 0.05 an agent's decision would appear reversed to both itself and its neighbours – and hence all misremember it. As a result performance gets slightly worse, and the best value for the chance of learning from others now appears to be higher. Indeed, if learning chance is set to 1, the introduction of noise actually improves performance. With noise, agents appear to be making decisions that are surprising in the light of past history. So observing agents record the actions, and have less certainty concerning which behaviour history is recommending to them. Because of the reduced certainty, the next decision maker involves its private signal in the decision, which, as long as it too is not reversed by noise, is likely to reflect the current bias. So, surprising actions can be of benefit to the population as a whole, even when they stem from random processes rather than from some genuine information source about the world.

There is plenty of scope for variations on this model. For example, the agents do not know whether their neighbours are interconnected. Agents with more knowledge about the local social network could factor this into their decision making, as they could observations of past correlations between neighbours' decisions. In addition, agents could hold opinions about their neighbours, including what type of person that neighbour is, and how such a type might differ from oneself in his or her assessment of the same evidence. If decisions are being made on multiple issues, evidence of decisions on one issue might guide one in evaluating neighbours' decisions on some new issue – for example, a neighbour whose decisions tend to correlate with one's own might be expected to correlate again in the future. Acemoglu et al. (2011) come to related conclusions about the value of social network structure for preventing information cascades, but they use Bayesian analysis where we use agent-based simulation.

To conclude then, adoption decisions need to be based on both information about other people's decisions and one's own private sources, although the ideal balance is more towards the former than the latter. However, what the ideal balance is turns out to be yet another phenomenon that varies with social network structure. Indeed, thanks to our choice of others to learn from being constrained by our social networks, a population can avoid the problems of fads illustrated in the information cascades model.

CONCLUSIONS

With the aid of simulation models we have seen how the structure of social networks can affect diffusion outcomes, with implications for processes that depend on these outcomes, such as collective learning and problem solving.

We examined familiar properties of social networks, such as their density, the degrees of separation between their nodes and the existence of clustering patterns. The denser a network is with links, the more likely it is that an innovation introduced by one agent can spread to the others by such communication as word-of-mouth advertising and imitation. Density also shortens the paths between nodes by which the innovation must travel.

Patterns among the links matter. The diffusion success of competing innovations can depend on the network attributes of the nodes at which the diffusions begin. When a firm can, through targeting its advertising efforts, influence with which node diffusion begins, then awareness of the nodes' metrics would have practical implications. But the extent to which different node metrics predict diffusion success varies with the overall network structure. For example, the tree-like Barabási–Albert scale-free network calls for a different strategy than that for networks with some degree of clustering. Clustering of nodes can encourage groupthink, when a group well-connected within itself locks into a particular decision – to adopt or not – and is then resistant to change and influences beyond the cluster.

The focus in this book is on simulation models rather than the analysis of empirical data, but the prospect was discussed of obtaining data on real social networks with which to validate agent-based models' use of network structures. Research into the networks formed by electronic communications has grown quickly in the age of the Internet, but many of our most important social relations are based on interactions conducted face-to-face, not electronically. It remains to be seen whether technology can provide us with better data on the relations that matter most.

Whatever the true network structure, diffusion processes exhibit path dependence when rival innovations are diffusing. Relative success breeds more success. But network structure can influence the question of which path-dependent outcome emerges, with some regular structures giving more chance to a fair balance between rival innovations.

Lock-ins and path-dependent behaviour have implications for problem-solving performance, and initiatives that modify social network structure may lead to improvements in this. Too much sharing of information can also lead to herd behaviour and fads, when decision makers all react to the same information about adoption. However, in the context of the information cascades model it was shown that the constraints placed on interaction by a social network structure can remove this danger, and collectively populations in social networks can track changing knowledge well.

The simulation models all point to interesting areas for investigation, and we hope one day empirical datasets on real social networks will be tied in with these. In addition, data on the physical and electronic networks that constrain social networks, such as street grids, office layouts and the Internet and World Wide Web, may help us understand how cities, organisations and now the Internet influence innovation generation and diffusion. Considering the variety of network structures, their implications for diffusion and their dynamics, it will be apparent that studies of the diffusion of innovations need tools that can incorporate all these, which, of course, agent-based simulation models can.

4.   Explore and exploit

INNOVATION IN ORGANISATIONS

At the end of the previous chapter a model was discussed in which agents needed the diffusion of new information for a purpose: keeping track of their environment. This was a task which they could perform better collectively, via their social network, than they could if they acted independently. But they were vulnerable to herd behaviour, collective mistakes that individuals struggled to correct. This chapter continues the themes of innovation being useful for some task or end, and of agents' behavioural practices and social network structure having an effect on their ability to perform that task. In particular, performance depends on a balance between *exploring* new ideas, or generating innovations, and *exploiting* innovations already found, diffusing them to others. The next chapter will cover models of scientists exploring and exploiting ideas through academic publications. This chapter examines the use of simulation models of the ways in which organisations and the people working in them innovate in their day-to-day activities, what are called *models of organisational learning*.

Much of the work on organisational learning draws upon what has been called the 'Carnegie School', a tradition, originating out of what was then called Carnegie Tech in the United States, of thinking about organisations as engaged in problem solving (Argote and Greve, 2007; Gavetti, Levinthal and Ocasio, 2007). Success in problem solving is deemed to be influenced by the behavioural practices of the organisation's members and the structures of their interactions. Experimenting with these in real organisations is often impossible, but a computer simulation of organisations allows one to try out alternative organisational structures and practices on some simulated problems, with a view to understanding why real organisations have the problem-solving performance they have, and perhaps also making recommendations about changes to structures and practices.

This chapter highlights several issues for modellers of organisational problem solving. One key issue is the representation of the task or problems to which solutions are sought. Another issue is the representation of agents' goals and how rewards for problem-solving knowledge are distributed through the organisation. In particular should agents be represented as seeking to improve their own beliefs, their organisation's official view, the average quality of belief among their organisation's members, or the quality level of the best belief among the organisation's members? Finally, we note how if the rewards for organisational problem solving are distributed in the manner of a winner-takes-all competition, competitors may have an incentive to take more risks in their search for better solutions, even if it means sacrificing some of their expected, or average, performance.

INNOVATION AS ORGANISATIONAL LEARNING

The 'Carnegie School' tradition in organisational studies began with psychologist Herbert Simon's critique of the representation of decision making agents in economists' models (Simon, 1955a, 1957). *Homo economicus*, whether a person or a firm, was deemed to be a rational chooser between the expected values of all available options, given perfect information about the probabilities, costs and benefits attached to those options, even when the models also assumed the numbers of agents and options were infinite. This assumption of rational agency had the advantage of being tractable to mathematical treatment under certain circumstances, and claims were made for the usefulness of its empirical predictions. But as a belief about how human beings make decisions it was clearly false, and Simon offered an alternative view. There are finite limits to humans' abilities to gather information and consider options, and rather than engage in time-consuming, expensive thinking, during which conditions might render some information obsolete, humans employ rules of thumb, or heuristics, to make rapid decisions. Unlike the economists' models, based on calculus, decisions based on heuristics are unlikely to discover the optimal solution to any problem, but they often reach a high standard that is close to the optimum, and 'good enough' for the decision maker's purposes. Thus in place of optimisers, Simon's agents are satisficers, seeking only satisfactory solutions. In place of perfect rationality, they are only boundedly rational. Later studies in psychology have born out this view of human agents as relying on heuristics (Gigerenzer, Todd and ABC Research Group, 1999; Kahneman, 2011; Kahneman et al., 1982; Klein, 2009).

Simon's behavioural theory was then applied to organisations or firms (Cyert et al., 1964; March and Simon, 1958), with Nelson and Winter (1982) combining it with ideas from evolutionary theory to revive evolutionary economics. Actions performed on behalf of organisations by their

Simulating innovation

members are chosen for their appropriateness in particular situations, according to various 'standard operating procedures' or 'routines' recognised by the organisation. Routines are based upon interpretations of past experience, and adapted in response to feedback about outcomes. The outcomes are evaluated in the light of the organisation's goals or targets, themselves often the result of struggle, bargaining and political negotiation between the organisation's members who may have a variety of conflicting personal goals (March and Simon, 1958). In summary, organisations learn, or adapt to fit better their environments, by being routine-based, history-dependent and goal-oriented (Levitt and March, 1988).

By the generic term 'routines' should be understood 'the forms, rules, procedures, conventions, strategies, and technologies around which organisations are constructed and through which they operate. It also includes the structure of beliefs, frameworks, paradigms, codes, cultures, and knowledge that buttress, elaborate, and contradict the formal routines.' (Levitt and March, 1988, p. 320) Partly they are encoded in the formal documentation of the organisation, its training courses, technical manuals, recruiting, promotion and accounts procedures, etc. But partly they are known tacitly by the organisation's members, and must be learnt by new members through either trial-and-error experience or through observation and imitation. Either way, routines represent knowledge that can survive turnover among individual members, and partly account for the organisation's own ability to exist beyond particular membership.

If 'routines' or 'standard operating procedures' sound like the 'procedures' executed as part of computer programming, it may come as no surprise that this tradition produced some of the first attempts, in Cyert et al. (1964), to use computer simulations to understand social phenomena. Nelson and Winter (1982) then introduced simulation to evolutionary economics. But it should be remembered that the inspiration for this tradition came from observations, or qualitative studies, made in real organisations (Argote and Greve, 2007)

For any particular routine, alternatives might exist, some of which may suit goals better than the current one, especially in an environment that is changing. Thus there is a need for innovation in routines. Such innovation is the result of search among the alternatives, for which the organisation may have more routines (meta-routines). March (1991) describes an agent-based simulation model of organisational learning in which two processes serve to adapt an organisation's coded routines in order to track its external environment. (A NetLogo version of this, *MarchOLM.nlogo*, is available on the website.) First, members of the organisation adapt their beliefs to resemble more closely the organisation's code through a process of 'socialisation'. Second, through a process of 'organisational learning' any members whose beliefs match reality overall better than the code does can revise the code with some of their views. Having superior knowledge gives members the self confidence to try to direct the organisation, but if there are multiple superior members with diverse beliefs, then as with goals in real organisations (March and Simon, 1958) any revisions to the code will be the result of conflict between these superior members, rather than a process that identified more accurately which of their beliefs actually match reality. March showed that the strengths or rates of these two processes, socialisation and organisational learning, could affect the success with which the organisation's code and the members' beliefs tracked reality. Agent members of the organisation began with diverse beliefs and were then the source for alternative views of reality. Rapid socialisation would reduce this diversity of opinion, but perhaps remove a correct view in favour of the official one in the organisation's code. Also, learning from the currently superior agents did not always involve updating the code with the beliefs that made them superior. If the code was replaced with false beliefs, socialisation could then lead to these being perpetuated. Workers needed to learn quickly, but not so quickly that good ideas were missed or replaced. In general, March argued, a balance was needed between *exploration* of new beliefs and *exploitation*, through socialisation, of existing ones. This applied also to dynamic environments. Knowledge could quickly become out of date when the environmental reality underwent a little 'turbulence', while the organisation converged on a homogeneous set of beliefs. However, a process that reintroduced variety to the population, such as 'turnover' in staff bringing in workers with new beliefs, could combine with learning to track the changed reality. In total the model organisation represented an evolutionary system, since it provided variation (via turnover), selection (via forming the group with superior beliefs) and retention (via learning and socialisation) (Rodan, 2005). Balancing exploration and exploitation means balancing variation and selection–retention.

This trade-off was not limited to computer-simulated organisations. (Levitt and March 1988). Very often, firms' growing familiarity and competency with some routines over others ('learning by doing') led to repetition of the familiar before the alternatives had been properly evaluated. In this 'competency trap' firms could converge prematurely on routines that were not the best. It might actually pay firms to maintain some degree of ambiguity or imprecision in their code and processes of transmission, so that variety was preserved for longer. This would be particularly valuable in a dynamic environment, not least because firms apply their routines in environments that include other firms: competitors, customers, suppliers and complimentary producers. In an ecosystem of interdependent organisations, one firm's innovation in behaviour may alter the environment of other firms, thus influencing the value of their current routines, and prompting them to revise these, which in turn affects others. Although firms have an incentive to make changes in order to adapt to their environment, learning from the changes is difficult when they can lead to a changing environment. Thus, superior performance is not guaranteed to follow either from learning, or from an incrementalist approach that emphasises frequent, multiple, small changes.

This raises the question of how best to search for innovative routines. Two methods were clear from studies of organisational learning (Levitt and March 1988): *trial-and-error experimentation* and *learning from others*. But how best should one go about generating new solutions to an organisation's problems, and how readily should one adopt innovations? March's distinction between exploration and exploitation has been much cited, and his computer model has seen some attempts at replication and extensions. (A special issue on March's paper appeared in the *Academy of Management Journal*, 49(4), 2006.) For example, Rodan (2005) adds trial-and-error experimentation by individual workers to compare with the learning they are capable of in the basic model simply by being part of a well-organised firm. Again the experimentation rate influences the balance between exploration and exploitation, although in a complex way when the environment is turbulent. However, March's model contains multiple processes that are stochastic (i.e. they involve random variation), and the model's output shows a high variability, making it difficult to obtain reliable summaries of the model's behaviour in experiments. Hence there was a need for alternative tools for thinking about organisational learning.

SEARCHING A LANDSCAPE

A later and more popular approach to models of organisational learning is based around a metaphor of searching for peaks on a landscape, using simple rules of thumb or heuristics.

This begins with the reflection that sometimes the consequences of our innovative actions are not foreseeable in advance. We just have to try a particular action, and see if it makes things better or worse. How effective is this as a way of solving problems? When might trial-and-error be a good strategy for problem solving?

Suppose you are walking on a hill or mountain in thick fog. Somewhere on this landscape there are some mountain peaks, but the fog prevents you from seeing one pace in front of you. How can you find a peak? Fortunately, if you take a step forwards or backwards, you are able to

Figure 4.1   A simple landscape with two peaks. Two agents stand ready to try to find the peaks.

judge whether you have just gone uphill or downhill. This means the following rule of thumb, *random-walk hill climbing*, could find you a peak:

Repeat
        Trial: Take a step in a randomly chosen direction.
        Error?: If this step takes you downhill, go back to where you just were.
Until surrounded by downhill steps (or just tired)

This rule of thumb, or heuristic, is an example of 'local search'. A search agent takes small steps each time, and cannot leap across large sections of landscape. On the other hand, there are relatively few possible small steps available from the agent's current position to choose between. In the one-dimensional landscape of Figure 4.1, each agent has just two moves ('left' or 'right') to choose between. While on flat ground, the agent will be indifferent between the moves it makes – the walk will be truly random. But once an agent reaches the slope of hill, as it is likely to eventually, then the agent's preference for keeping uphill moves mean that very soon it progresses up the hill to the peak. Once at the peak, all paths lead downhill and so are not taken. In terms of problem solving, an optimal solution – in this case, a position with a peak height – has been found.

Figure 4.1 shows a landscape with *two* peaks, however: one taller than the other. How do we know our search agent will find the tallest peak in the landscape, the 'global optimum'? As a heuristic method, random-walk hill climbing tends to find peaks quite efficiently, but it offers no guarantee that the peak it finds will be particularly tall. In Figure 4.2, two search agents have ended up on different peaks, one taller than the other.

Another feature of the landscape depicted is that one peak is broader than the other. If many agents were 'dropped' onto this landscape at random locations, approximately half of them would land on the right-hand hill, thus increasing the chance that they discover its peak rather than the one on the left. Indeed, any agents landing to the right of the right-hand peak have no chance of ever reaching the left-hand hill, since to reach it they would have to overcome the right-hand peak. This highlights

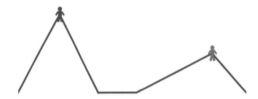

Figure 4.2   Different searches can reach different peaks, with no guarantee of finding the tallest.

a potential problem with the random-walk hill climbing heuristic: that search agents can get stuck with inferior solutions, usually one near to their starting point, a so-called *local optimum*.

One variation on the method, *simulated annealing*, allows agents to take downhill steps sometimes, based on a metaphor with the physics of heated particles on a rough surface (Kirkpatrick, Gelatt and Vecchi, 1983). Steps uphill are always accepted. Steps downhill are accepted with a chance that depends partly on the size of change in height, with large drops in height being less likely to be accepted. The chance also depends on a scalar variable, called 'temperature', which decays over time, making large drops less and less likely as it does so. Gradually the search method becomes indistinguishable from hill climbing. The agent is then stuck with whichever peaks it happens to be between. However, for some landscapes, there is a good chance that the earlier random wandering took it into an area with taller peaks than that it started in.

Of course, the example landscape in Figure 4.1 is simplistic. There is just one decision to make, i.e. one position to choose, and a relatively easy-to-explore two-peak evaluation of possible positions. This was enough to communicate the metaphor of search on a landscape, but seems too simple to tell us much about the benefits of different search practices. Real organisational decision making involves making multiple decisions, and may involve much more difficult search tasks.

KAUFFMAN'S NK FITNESS LANDSCAPE

What are needed are models of more challenging landscapes. Theoretical biologist Stuart Kauffman introduced a design for a fitness landscape that is relatively simple to describe and compute, and has been used in many simulation models. His *NK fitness landscape* was devised as an analogy for the search task faced during biological evolution, when combinations of genes for organisms that fit their environments well are sought by natural selection (Kauffman, 1993, 1995). Some genes are known to affect the role of other genes, and these interdependencies turn genetic evolution into a non-trivial search problem. The formal model has some similarity to spin glass models in physics, where one talks of finding positions of minimal energy rather than genes of maximal fitness. Kauffman has since extended the model to include co-adaptation between different species (the *SNKC* model) (Kauffman, 2000), and variations on it have been applied to strategic management and organisational learning (Caldart and Oliveira, 2010; Lazer and Friedman, 2007; Levinthal, 1997; Rivkin and Siggelkow, 2003; Robertson and Caldart, 2008). Its application to models of technological innovation (Frenken, 2001, 2006a, b) will be mentioned in Chapter 7. In this section we describe a basic version of it for use in models of collective learning.

Solutions are sought to problems consisting of several interdependent parts. These parts can be thought of as decisions to be taken about actions to perform or traits to adopt. Each decision making agent, a person, firm or species, faces N decisions of a simple yes/no or true/false form, and their current response, or solution, is represented as a string of N binary digits, or bits. Each decision variable contributes to the fitness value of an agent's current solution, and contributions are taken from tables of fitness values, one table for each of the N variables. However, each table has multiple rows. Which row is read depends on the current state of the corresponding decision variable, plus K other decision variables belonging to that agent. Thus, each variable's table has 2^{K+1} rows of fitness contributions. When a fitness contribution has been calculated for each decision, the contributions are averaged to produce the fitness value of the whole solution.

During simulation initialisation, random sampling decides for each variable which K other variables to read from when choosing a row of its fitness table and what the fitness contributions are in the table. Both of these are then fixed for the rest of the simulation run. Fitness contribution values are generated by sampling from a continuous random variable uniformly distributed over the range 0 to 1, or U[0,1). Each variable's K input variables are chosen from the agent's other variables with uniform preference. This creates a random network of interdependencies between the variables, with each variable having K inputs besides itself, and on average K outputs besides itself. Table 4.1 shows inputs and fitness contribution tables for an example landscape with $N = 4$ and $K = 1$.

After initialisation, a simulation run consists of repeated performance of heuristic search steps, usually for a fixed number of iterations or until a peak seems to have been reached and there is no improvement in the fitness of solutions. The heuristic search steps involve changing an agent's solution, the set of its decision variable values. Kauffman (1993, 1995) initially used a random-walk hill-climbing algorithm to guide these changes. That is, each time step, an agent chooses one variable at random and

Table 4.1   *Example fitness tables for an* NK *landscape, with* N = 4
             variables, and K + 1 = 2 *input variables to create the*
             interdependencies between variables' fitness contributions.
             Each variable's table has 2^{K+1} = 4 *rows, one for each possible*
             combination of input values.

Variable 0			Variable 1		
Input variables:		Fitness	Input variables:		Fitness
0	3	contrib.	1	0	contrib.
0	0	0.852	0	0	0.799
1	0	0.643	1	0	0.300
0	1	0.599	0	1	0.261
1	1	0.422	1	1	0.646
Variable 2			Variable 3		
Input variables:		Fitness	Input variables:		Fitness
2	1	contrib.	3	2	contrib.
0	0	0.575	0	0	0.707
1	0	0.623	1	0	0.854
0	1	0.984	0	1	0.881
1	1	0.825	1	1	0.029

reverses its value, from 0 to 1 or from 1 to 0. Fitness is then recalculated
for the resulting solution. If the fitness value is not worse than that for the
previous solution, the new solution is kept. Otherwise, the decision change
is reversed and the old solution and fitness value returned to.

When an agent changes one decision, this will typically alter the fitness
contribution made by that variable, but also the fitness contributions made
by any variables for which the changed one is an input. Given the random
network of interdependencies defined at the start, on average $K + 1$ fitness
contributions will be changed. It is these interdependencies between variables that make for interesting search problems. In the simple fitness landscape in the previous section there were just two peaks, or optima, to find.
The possible combinations of variable values, or solutions, can also be
thought of as a landscape, one made up of locations (solutions) connected
by paths to those other locations that differ from them in one and only
one variable value. An example solution space, or fitness landscape, can
be seen in Figure 4.3, based on the same inputs and fitness contributions
defined in the tables of Table 4.1, and generated by our NetLogo program,
SolutionNet_KauffmanNK.nlogo, available on the website.

An *NK* fitness landscape gains peaks as the number of interdependen-

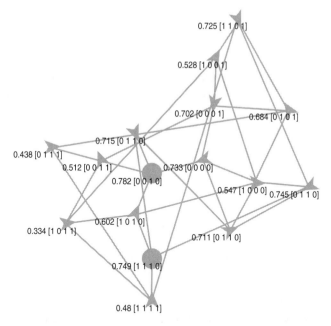

0.725 [1 1 0 1]

0.528 [1 0 0 1]

0.702 [0 0 0 1]   0.684 [0 1 0 1]

0.715 [0 1 1 0]

0.438 [0 1 1 1]

0.512 [0 0 1 1]   0.733 [0 0 0 0]

0.782 [0 0 1 0]

0.547 [1 0 0 0] 0.745 [0 1 1 0]

0.334 [1 0 1 1]

0.602 [1 0 1 0]

0.711 [0 1 1 0]

0.749 [1 1 1 0]

0.48 [1 1 1 1]

Figure 4.3   *The* NK *fitness landscape defined in Table 4.1 depicted as a network of solutions, or solution space. Here* N = 4, *giving 16 possible solutions, and* K = 1. *Each solution is linked to all other solutions that differ from it by one and only one value. Solution labels contain the combination of variable values the solution represents and its associated fitness value. There are two peak solutions, or optima, shown as circles. The other solutions are shown pointing towards their best neighbouring solution.*

cies, K, increases. The more peaks there are, the easier it is to move up one, but the less chance there is that the peak climbed will be the tallest peak on the landscape. Until an agent visits a peak, it has no means of knowing how tall it is. So finding one of the taller peaks becomes harder as K increases. While K remains very much lower than N, however, there is still considerable correlation between the fitness values of solutions at two adjacent time steps, since these solutions differ only in one bit, and their fitness values are averages of sets of N fitness contributions that differ only in approximately $K + 1$ elements. It is this correlation that makes *local search* a useful method, i.e. stepping from one solution to another by changing one bit at a time. Consider trying to explore a highly rugged landscape. The height at one's current position might give no guidance as to the heights of neighbouring positions, and whether a step to the left was going to plunge one down a deep ravine or offer a climb to a tall peak. A virtue of Kauffman's model is that the *ruggedness* of the landscapes, and hence the difficulty of searching them, is tuneable by controlling a single parameter, K.

As a model of the tasks faced during decision making and organisational learning, the NK model has a few drawbacks and has prompted some variations (Robertson and Caldart, 2008). The problem-solving agent is highly simplified. It remembers nothing about its past history other than its most recent solution, and does not build any overview of the landscape it has explored. Neither does it reflect on its success to date and revise its search method. In addition, it does not look forwards. Extensions to the model include giving agents cognitive representations of strategic plans to guide their search processes (Gavetti and Levinthal, 2000), and giving them powers of reasoning by analogy to past experience (Gavetti, Levinthal and Rivkin, 2005). Decisions in the basic model can be taken in any order, they are homogeneous in the number of dependencies they have (K), and their contributions to fitness carry equal weight. In organisational decision making, a hierarchy often applies, with higher level decisions constraining lower level ones (Gavetti, 2005). Some decision areas are considered more important to the firm than others, with some identified as 'core' and others 'peripheral', and even within core activities such as research, development, human resources and advertising, decisions differ in their scale and timing of impact. Perhaps the most problematic feature of the NK idea is that the fitness landscape it describes is static: it is fixed at initialisation and unresponsive to the actions of the search agent. Real organisations can expect to affect their physical and economic environments to some extent, and also expect them to change over time due to factors outside the organisation (Siggelkow and Rivkin, 2005).

TASK DECOMPOSABILITY AND INTERDEPENDENT AGENTS

Kauffman himself has provided a special case of landscape dynamics, by introducing multiple agents, each of which has a fitness landscape coupled to some of the landscapes of the others (Kauffman, 2000). Kauffman introduced his *SNKC* landscape model as a representation of multiple species co-evolving. The S stood for species, but when discussing innovation it makes more sense to think about competing firms (Caldart and Oliveira, 2010) or interdependent individual workers in an organisation (see below). The *SNKC* concept also allows us to represent an important feature of real tasks. Herbert Simon (1962) noted that complex systems encountered both in biological and social studies are usually modular, consisting of

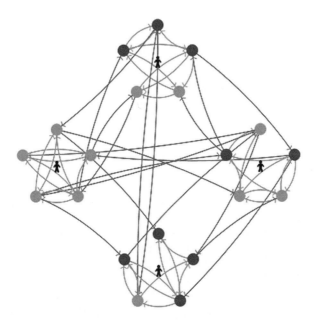

Figure 4.4   An example SNKC *network.* S = 4 *agents each have* N = 5
*decision variables, shown as nodes. Each node has links to take
inputs from* K = 2 *nodes owned by the same agent, and from*
C = 1 *node owned by another agent.*

multiple subunits. The design tasks represented by these systems can typically be decomposed into more manageable subtasks. What makes it possible to decompose a task is that interdependencies between the subtasks
are far fewer in number and importance than those within the subtasks.
The NetLogo program *NKNetworks.nlogo* on the website generates such
SNKC models and simulates searches using a choice of methods, including
random-walk hill climbing.

Consider a system to include multiple agents. Each agent has its own
fitness landscape, but each decision variable in its solution now takes input
from not only itself and *K* other variables belonging to that agent, but also
another *C* variables belonging to the solutions of other agents. Put another
way, the value of an agent's decision depends not only on its other decisions
but also on other agents' decisions. This means that a variable's fitness
contribution is now taken from a table with $2^{(K + C + 1)}$ rows. As with *K*, the
behaviour of the system as a whole can be tuned by altering the parameter
C. Figure 4.4 shows an example network of four agents, each having *N* = 5
decision variables, and each variable taking *K* = 2 inputs from other variables belonging to the same agent, and *C* = 1 input from variables belonging

Simulating innovation

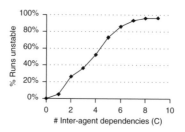

Figure 4.5   *As the number of inter-agent dependencies,* C, *increases, 2000-step simulation runs become less likely to reach stable system states, where all agents have peak solutions.*

to another agent. The effect of these latter interdependencies is to make the agents' search problems interdependent. One agent can take a step to improve its own fitness value, but in making changes it will alter an input to another agent, and may thus alter that second agent's fitness landscape. Too many such interdependencies (higher values of C) will mean agents' landscapes deform too fast for any of them to make much progress in improving their own fitness. They may, as Lewis Carroll's Red Queen explains to Alice, have to run fast simply in order to stay in the same place (now known as the *Red Queen Effect*), though in fact even keeping one's fitness is impossible to guarantee in such a system. Figure 4.5 shows for four-agent systems, with $N = 5$ and $K = 0$, how increasing the inter-agent dependencies (C) makes it more unlikely that, after 2000 time steps, the system will have stabilised, that is, that all agents will have reached (local) optima or peaks.

In a multi-agent *SNKC* model agents can perform their searches ignorant of each other's activities, even though the fitness values of their resultant decision sets are coupled. But what if agents could incorporate information about each other into their searches?

One way in which this might happen would be if the effects of one agent's decisions on other agents' fitness were known to all. This opens up the possibility of agents performing searches for decisions that improve their total or collective fitness rather than that of the current search agent. Figure 4.6 shows that interdependencies between agents' landscapes can make searching for collective fitness (called here 'Altruism') a better strategy for the group as a whole than searching purely to improve one's own fitness ('Egoism'). Mean results are shown from 200 runs, with 95 per cent confidence intervals for each mean, using $S = 4$ agents, $N = 5$ and $K = 0$.

However, agents improving group fitness may produce a set of solutions that, while representing a peak for *collective* search, do not represent peak solutions for every member of the group. If, having reached this point from pursuing group fitness, one of the agents now switched to consider-

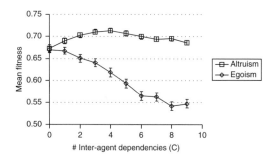

Figure 4.6   The more dependencies there are between agents, the greater the advantage in each agent trying to improve the group's fitness ('Altruism') rather than their own ('Egoism').

ing only its own fitness, then it might find it desirable to change decisions, which, of course, would impact on the fitness values of others in the group. In economists' terms, the peak the group has reached is not necessarily a *Nash equilibrium*; that is, some agents may be able to benefit from defecting from pursuing the group goal. One can test this in *NKNetworks. nlogo* by pausing the simulation after it has reached a stable set of agents' decisions via altruistic fitness improvements, then changing to egoistic fitness improvements and restarting the simulation run. In the resulting instability the group mean fitness is likely to fall back, and individuals may not get to keep their advantages from defecting (Figure 4.7).

So, agents can benefit in the longer term if they give up their short-term, selfish interests for the sake of a group of interdependent others. This indi-

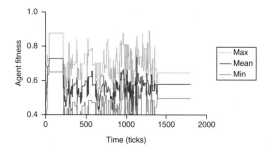

Figure 4.7   In this example with four agents, N = 5, K = 0, and C = 4, the system stabilised after the agents had pursued group fitness for just 40 ticks. At 210 ticks the agents' method was set to pursuing individual fitness. The system stabilised again after 1400 ticks, but at lower fitness.

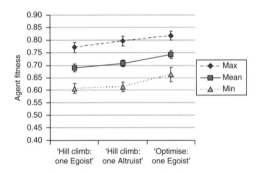

Figure 4.8   When task decomposition works: Max, Mean and Min Agent
fitness in a population of four agents, with N = 5, K = 3 and
C = 1. Three different search methods are compared, with
95 per cent confidence intervals for mean results, each run
including 6400 fitness calculations.

cates the need for members of a group to observe the actions of their fellow members and assess whether all are still pursuing the group fitness, and instigate punishment of any defectors in order to reverse any perceived fitness gains to the individual defector.

   In these examples, dependencies are between variables belonging to different agents (i.e. $C > 0$). What if some the dependencies are between variables belonging to the same agent (i.e. $K > 0$)? If $C = 0$, then we have agents whose decisions do not affect each other. The system may be decomposed into separate NK models, one for each agent, which then adopts an egoist method. In addition, each agent is solving a relatively small problem. $N = 5$ means there are $2^5 = 32$ possible solutions for each agent, and calculating the fitness of each solution is not unreasonable. By contrast, an NK model with the same number of variables as our four agents have collectively, i.e. $N = 4 \times 5 = 20$, would have $2^{20} = 1048576$ solutions. When an $N = 20$ problem can be broken down into four $N = 5$ problems, exhaustive search becomes feasible. This *decomposition* of a task into subtasks may also be possible when there is a small amount of coupling between the $N = 5$ subtasks. Figure 4.8 shows that when $K = 3$ and $C = 1$, the best search method is indeed to allow the agents to search every one of their possible solutions for their optimum. In this experiment, hill-climbing methods have been run for 6400 steps so as to involve the same number of fitness calculations as performing exhaustive search (of 32 solutions) for 200 steps. By contrast, Figure 4.9 shows that when the interdependencies are mostly inter-agent ($K = 1$, $C = 3$) rather than intra-agent, optimising at subtask level is worse than altruistic hill climbing.

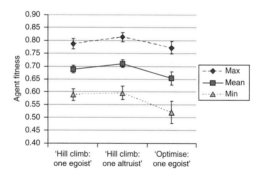

Figure 4.9 When altruism works: Max, Min and Mean Agent fitness in a population of four agents, with N = 5, K = 1 and C = 3. Three different search methods are compared, each equivalent to 6400 fitness calculations.

Drawing analogy with these models leads one to suspect that differences in innovation and learning behaviour between organisations, countries and industries may reflect differences in the task structure they face. When we are 'all in it together', with high interdependency, altruism and collaboration make sense, even though thinking about each other's actions and fitness makes for a larger problem. Under these circumstances, optimisation is infeasible, and we require heuristic methods of solution. When our fates are only loosely coupled, it is best to concentrate on one's own decisions and fitness, and use the search approach that offers the best returns. Under these latter circumstances, optimisation may prove better than heuristics.

Before the analogy can be relied upon, however, some limitations of the model as a representation of real organisations must be admitted (Robertson and Caldart, 2008). In Kauffman's *SNKC* model all agents have the same numbers of decisions and interdependencies (*N*, *K* and *C*). There is no distinction between competitive and cooperative interactions. Agents as firms are not divided into types, such as competitors, suppliers, customers, regulators and providers of complementary services. Also agents are identical in size and weighting when calculating the system average, but some interconnections between real firms may be more important than others. Kauffman's binary decision network models demonstrate the abstract concepts of interdependency and decomposability, and future research will have to investigate whether the analogy with real organisational problem solving can be improved.

LAZER AND FRIEDMAN'S MODEL OF COLLECTIVE LEARNING

With these caveats about current fitness landscape models in place, one can still use them to demonstrate properties of collective learning. In the previous section agents searched their own (coupled) landscapes, but when altruistic they could take each other's fitness into consideration. Another way in which agents might incorporate information about each other is if they are actually tackling the same search problems, that is, exploring the same fitness landscape. In such a case, they might benefit from learning of each other's solutions. The question then becomes how best to share information. When should one learn from others and when should one just explore on one's own? Lazer and Friedman (2007) have proposed a model of collective learning in an organisation that draws upon a version of Kauffman's *NK* fitness landscape.

Imagine an organisation of 100 agent workers. Each is concerned with the same complex problem, represented as $N = 20$ decisions. Lazer and Friedman evaluate every agent's solution set using the same fitness definition, a moderately difficult *NK* landscape with $K = 5$ interdependencies per decision, giving $2^{20} = 1048576$ possible solutions to choose between, and several hundred local peaks on the landscape to get stuck on. It is these that make learning from others an advantage. From an initial starting point an agent can climb one of the nearest peaks but has no guarantee that this local optimum is among the tallest peaks in the landscape. Other agents who have started from different points will mostly discover different peaks, some of which may be taller. Hence, an agent benefits from consulting other agents to see if any have better solutions, and imitating one of those that does.

The website contains a replication of Lazer and Friedman's model as an Excel/VBA workbook, *LazerFriedmanModel.xls*. The main algorithm of the simulation is as follows:

```
Initialise NK landscape
# Initialise population of agents:
For each agent
        Set current-solution to a randomly generated solution.
        Set current-fitness to fitness of current-solution
Next agent
For each time step:
        For each of the agents
                Set changing flag to FALSE
                # Stage 1 Learning from others:
```

If there are any agents in my neighbourhood with *current-fitness* greater than my
current-fitness,
> Set *Alter* to be one of my fittest
> neighbours.
> Set *next-solution* to be an
> (potentially imperfect) copy of *Alter's*
> *current-solution*.
> If the fitness of the resulting
> *next-solution*
> <= *current-fitness*,
> > Set *changing* flag to TRUE

\# Stage 2 Trial-and-Error Experimentation:
> If *changing* flag = FALSE,
> > Set *next-solution* to be a perfect
> > copy of *current-solution*.
> > Pick a random bit in *next-solution*
> > and reverse its value.
> > If fitness of *next-solution* >
> > *current-fitness*,
> > > Set *changing* flag to TRUE

Next agent
For each of the agents
> If *changing* = TRUE,
> > Set *current-solution* = *next-solution*.
> > Set *current-fitness* = fitness of *next-solution*

Next agent
Output periodically the population mean and max fitness.
Next time step

The performance of the organisation under this algorithm depends on
how we define an agent's 'neighbours'. Suppose every agent is able to learn
from every other agent. That is, the agents in the organisation interact in a
complete network. Suppose also that at initialisation, all agents are unique in
solutions and in fitness values (which is highly probable). In that case, there
will be one agent who has the best solution in the organisation. In the first
time step, or iteration of problem solving, that agent will be unable to learn
from anyone else, since no one else has a superior solution. So that agent
will employ trial-and-error experimentation only. All other agents will copy
its solution. If the first agent's experimentation was successful, that agent
will have a new, improved solution in the second time step. All other agents
will then be able to imitate that. However, at some point this leading agent

is likely to fail to find an improvement. At this time step, the rest of the population will catch it up and all adopt the last improved solution it had. In the next time step the population will be homogeneous, and thus no agent will stand out as having superior fitness to any other. Whereupon all agents will skip learning from others and employ experimentation instead. If any of the $N - 1 = 19$ variations on their shared solution are improvements, it is highly likely that at least one agent in a population of 100 agents will find one. The population will then return to the phase of the best agent using trial and error, and the other 99 copying it. Thus, although the population has 100 agents, capable of working on 100 distinct solutions, from the first time step onwards, in effect the organisation is exploring variations around a single path. This path began at the then best solution in the initial population. But the initial solutions' fitness values turn out to have negligible correlation with the fitness values of the final solutions reached. Converging on the early best solutions is not the best route to good final solutions. With a complete network among the organisation's members, there is a high chance of premature convergence and finding only a relatively inferior peak.

Contrast this scenario with one in which agents are organised in a *linear network*, that is, a line of agents, where each agent can interact with just two neighbours, except for the agents at the ends of the line, who have just one neighbour. Agents make comparisons with far fewer neighbours than before, so there is more chance for them to be leaders, engaged in trial and error, rather than followers, learning from others. Thus the linear-network organisation maintains rival search paths for much longer than the complete network did. On the other hand, when an agent finds a good solution, it takes longer for this to diffuse among the other agents. If the agent on the end of the line of 100 nodes finds a global optimum, it will take 99 iterations for it to reach the agent at the other end.

Figure 4.10 compares these two network architectures with a third, the von Neumann architecture, that is, a regular, four-neighbour two-dimensional grid network with wrap-around edges. With each choice of network, simulation runs were performed on 1000 fitness landscapes. Using a common set of 1000 fitness landscape problems the three different network architectures were tested over 50 iterations. The chart shows averages for the 1000 problems, output at 10 points in time, or every five iterations. It repeats Lazer and Friedman's finding that the complete network converges fastest, and if only 10 iterations were run this would be the best architecture. However, with more iterations, architectures that converge more slowly are able to explore more of the landscapes, and thus converge on taller peaks. Forty iterations seem enough for the linear network to overtake the four-neighbour grid.

If the agents are arranged in an Erdős–Rényi random network, introduced

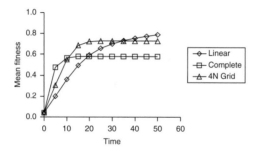

Figure 4.10   *As Lazer and Friedman (2007) showed, network structure can matter. Here a complete network converges more quickly, but reaches worse peak solutions than can be obtained by a linear network or a four-neighbour grid given sufficient numbers of iterations.*

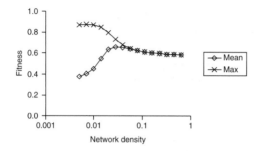

Figure 4.11   *The mean and maximum fitness obtained vary with the density of the agents' social network when this takes the form of an Erdős–Rényi random network.*

in the previous chapter, problem-solving performance can be measured for various levels of network density. As Figure 4.11 shows, there exists a density at which the organisation's mean agent fitness peaks. At lower densities the network tends to be fragmented into disconnected components, and agents in one component will be unable to copy solutions found in another component. One the other hand, the fragmentation makes it highly likely that the separate components will converge on distinct optima, some of which may be much higher than the others. Hence, the best fitness value in the population of agents ('Max') improves as density is decreased, though further investigation reveals that the best density for maximum fitness is greater than 0, so at least some learning from others is still desirable.

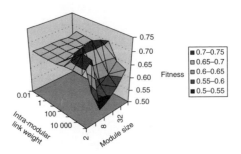

Figure 4.12   *Mean fitness in a modular network, with varying numbers*
              of agents per module ('module size') and varying degree of
              preference for distributing links within modules rather than
              between them ('intra-modular link weight').

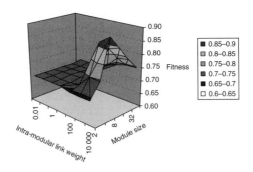

Figure 4.13   *The maximum fitness in the population in the same networks*
              as before.

The number of links in the network is not the only consideration. How those links are distributed is also important. In a Figure 4.12 a population of 128 agents has been allocated to various numbers of groups or modules. A fixed number of links (representing a density of $1/32 = 0.03125$) are distributed among the nodes, either to pairs of nodes within the same module (intra-modular links) or to pairs of nodes in different modules (inter-modular links). Node pairs are selected using sampling stratified by a weight on intra-modular pairings which can be varied. If this weight is high, the resulting network is fragmented for all but the smallest of module sizes, to the detriment of mean fitness. Fragmentation leads to better values for the maximum fitness in the population, however (Figure 4.13). At lower values of the weight, the population converges on inferior peaks,

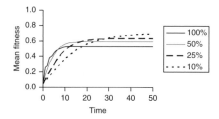

Figure 4.14   *Slowing down the rate of learning from others, using the*
Velocity *parameter, can lead to better mean fitness in
the organisation, if given sufficient time. Here learning
from others only 10 per cent of the time leads to better
performance than 25 per cent, 50 per cent and 100 per cent,
but it takes 30 time steps for the advantage to emerge.*

with population mean and maximum fitness equal. The ideal weight for
mean fitness varies with module size.

The advantage of slowing convergence can be seen in ways other than
comparing network structures. Lazer and Friedman introduce two variations on the basic algorithm that do this. First, they introduce a parameter
called *Velocity*, ranging from 0 to 1. This is the chance that the agent
performs stage 1, learning from others. The default value is 1. If set to 0,
agents always skip to stage 2, trial and error, and the population becomes
devoted to the hill climbing heuristic. Figure 4.14 shows experiments with
different values of *Velocity*. The second parameter, *Error*, also ranging
from 0 to 1, is the chance during the learning-from-others process of the
imitating agent failing to copy the imitated agent's bit and instead keeping
its own bit. The default value is 0 (no error, all bits copied). If set to 1, all
bits are kept and the effect is equivalent to skipping straight to stage 2.
Figure 4.15 shows the results on population mean and maximum fitness
for different values of *Error*.

Get the right error rate and the gap between linear and complete networks can be closed. Alternatively, we can say that for a given rate of
information transfer during imitation, some network structures are better
than others. Either way, though, you will have to choose between generating a population with the best solution (i.e. the best value for maximum
fitness), and a population with the best mean fitness.

Do you want the best solution or the best mean solution? One might think
it easy: obtain the best solution; then everyone else can copy it so that the
population maximum becomes the mean as well. But adopting someone
else's solution is not always straightforward. People do not always want
to adopt an entirely different way of doing things. They may be resistant

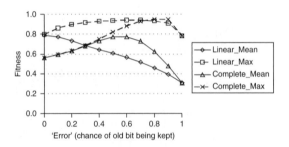

Figure 4.15   Network structure interacts with how much imitation occurs during an interaction.

to anything 'not invented here', or something in the development of which they feel they have played no part. It may also be too costly or slow to adopt all of the new solution, and so one is forced to introduce it in stages, which may lead to many intermediate solutions that are neither as good as the optimum nor as good as the solution currently held by those one wants to convert. Thus, practices which involved more people in the development of the final solution, and thus gave more of them a feeling of ownership or responsibility over it, might prove more successful for the organisation as a whole than those which assume everyone will accept the best solution simply because it is the best.

We can simulate this preference for similar solutions as follows. When agents have selected a neighbour with the best fitness, let them try to copy the selected neighbour only if the selected neighbour's solution matches their own by some given proportion. If this *similarity threshold* is not crossed, then agents skip on to trial-and-error experimentation instead. Figure 4.16 shows the interaction between *Error* and *Similarity Threshold* for 200-iteration runs on a complete network. When *Error* = 0 (the original value) and agents copy all 20 bits of neighbours' solutions, the best value fitness is obtained with a similarity threshold of around 0.5, equivalent to 10 bits. At lower values, convergence is too fast. At higher values, agents' ability to understand and learn from each other is too hindered, and only trial and error can result in improvements. However, this advantage due to similarity preference is dwarfed in scale by that to be had by increasing *Error*. At higher values of *Error*, the best similarity threshold is lower. Figure 4.17 shows that when it comes to the population maximum fitness rather than mean fitness, the interaction between *Error* and *Similarity Threshold* is different. At *Error* = 0, similarity preference actually helps performance. It prevents learning from others, and therefore convergence and reduction in the number of search paths. At high *Error*, however, a low similarity preference is better. The organisation can find a better

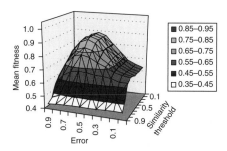

Figure 4.16   *Mean fitness varies both with* Error, *the chance of a bit not being copied a particular bit when learning from others, and* Similarity Threshold, *the proportion of a solution that must match an agent's own before they are prepared to copy it.*

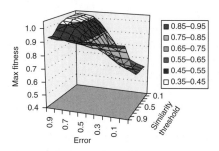

Figure 4.17   *'Max fitness', the best fitness achieved by any agent in the population, also varies with* Error *and* Similarity Threshold, *but peaks at different parts of the parameter space than mean fitness (Figure 4.16).*

solution by allowing agents to copy small parts of good solutions, potentially exploring exhaustively the variations around these to identify the best.

What if, instead of a preference for similar neighbours, agents preferred dissimilar ones, neighbours whose beliefs would seem more innovative? That is, if learning from others occurred only if the selected neighbour's solution matched an agent's in no more than some given proportion of bits? Experiments on varying this *dissimilarity threshold* produced charts very similar to those for the similarity threshold. Both factors interact with *Error*. Both show trade offs between mean fitness and maximum fitness. The best results in terms of mean and maximum fitness while varying *Dissimilarity Threshold* and *Error* are only slightly higher than those with *Similarity Threshold* and *Error*. What matters most, then, is slowing down

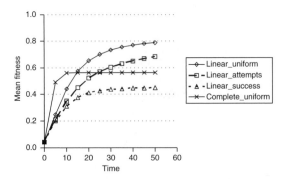

Figure 4.18   *Evolution in mean fitness for linear and complete networks when* Error = 0, *with opportunities to attempt to improve solutions distributed among agents uniformly, with preference for numbers of agents' past attempts, and with preference for agents' successful past attempts. The method for distributing opportunities mattered for the linear network but not for the complete network, so only the uniform method is shown for complete networks.*

convergence, however it is achieved, though some methods may be more effective than others.

One more type of problem-solving practice will be examined. So far, the Lazer-Friedman model has given every agent equal opportunity to try to improve its solution, whether by learning from others or by experimentation. But problem solving can require resources, and these can be distributed unequally. In particular, we often reward those who have previously been successful by giving them a greater share in future opportunities to seek improvements, some of which they may be successful in finding. This rich-get-richer principle, or cumulative advantage, will be returned to in Chapter 5 when we discuss the modelling of academic science. For now, we test it out in the context of the Lazer-Friedman model. In this version of the model, agents keep track of how many attempts they have made to improve their solutions, via learning from others or trial and error. They also keep track of how many of those attempts have been successful in producing an improved solution. These counts can then be used in stratified sampling for selecting agents to be given new opportunities to attempt to improve their solutions. Figure 4.18 shows results linear networks, with uniform preferences between agents (the original option), preference for agents on the basis of previous attempts at improvements, and preference for agents on the basis of previous successful attempts at improvements. It also shows results for complete networks with uniform preferences.

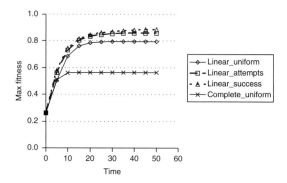

Figure 4.19   *'Max fitness', the best fitness in the population, also varies with the method for distributing improvement opportunities in a linear network, but the order is reversed from that for mean fitness. As with mean fitness, complete networks did not show any variation with distribution method, so only the uniform method is shown.*

When *Error* = 0, the impact on a complete network is negligible, and so the other options for complete networks are omitted from the charts. With the population converging so rapidly, which agent one selects to attempt improvements soon makes no difference. But for the linear network, the cumulative advantage processes can seriously harm problem-solving performance in terms of mean fitness. This harm applies also for complete networks with high values of *Error* (not shown here). In terms of population maximum fitness (Figure 4.19), however, giving preferential resourcing, whether on the basis of past resourcing or past successes, does yield slightly better results for linear networks, and for complete networks at higher *Error* (again not shown). Whether a policy of cumulative advantage makes sense for an organisation (or a society, for that matter) depends on whether it wants the best mean or the best maximum fitness. Of course, the organisation's policy may also have implications for how it remunerates its employees, and hence how well working for it satisfies their own goals, in which case a decision concerning cumulative advantage may affect staff turnover and other factors not included in this simulation model.

   In conclusion, then, Lazer and Friedman's (2007) original results and the extensions described here show how network structure and behavioural practices can interact to determine problem-solving performance in an organisation. A number of mechanisms were identified that could slow down convergence of the population's agents' search paths, and promote exploration of the landscapes. As well as social network structures restricting who could learn from whom, restrictions on the rate of learning, the amount of a solution copied, and the degrees of similarity and dissimilarity between interacting agents all had effects on convergence and thereby performance. In some cases, it would be possible to compensate for a change in one factor by altering another, such as changing network structure to compensate for restrictions on the amount of information copied during learning. We also saw that how problem-solving work was divided up among the organisation's agents could have an impact on performance. Giving extra resources to a few, such as the most successful to date, could undermine the mean performance of the population, but might yield a better solution for one agent than an even division of labour would. As with decisions concerning network structure and problem-solving practices, a manager has to decide whether the goal of organisational learning is a solution that all members have bought into and accepted as their own, or a solution that may be superior in external terms, but which only an elite have managed to reach. When the organisation interacts with its environment and competes with its competitors, which members of the organisation tend to express their beliefs and enact their solutions? Opportunities might tend to go to the members with the best solutions, or they might be more evenly spread. Different circumstances will call for different network structures and behavioural practices.

VARIATION IN PERFORMANCE

Models of organisational learning like those of March and Lazer and Friedman represent agents engaged in raising the organisation's value in knowledge or problem-solving fitness when performing routines, whether this value is some representative organisational code (as in March's model) or just the average fitness of all the member agents' beliefs. The aim of experimenting with such models is typically to identify ways of raising the *average* or *expected* performance of an organisation, both simulated organisations and real-world ones. One aspect of March's (1991) paper that has been less often remarked on is its section on how sometimes mean performance can be less important than *variance in performance*. Often in life we find ourselves competing in situations where only the top few participants will obtain any reward. Economists Frank and Cook (2010) list numerous examples of such *winner-take-all markets* in business, sport and the arts. In Chapter 5 we will see evidence of rich-get-richer processes, or cumulative advantage, in academic science. For now, we can illustrate them using March's example. (See the NetLogo program *MarchVariabilityDemo.nlogo* on the website.)

Consider a market in which there are N organisations plus your own, and each organisation's performance is a random variable. Suppose the N other organisations are identical in the probability distribution of their performances, for which we will use a Normal distribution with mean of 0 and variance of 1. Your own organisation's performance is also a random variable, and also normally distributed. But the parameters for this distribution, mean μ (*mu*) and variance σ^2 (*sigma squared*), will depend on the particular routines you employ. If your choice of routines yielded $\mu = 0$ and $\sigma^2 = 1$ then your organisation will have the same chance of achieving a particular level of performance as the other organisations. But what if the rewards in this market are not evenly distributed, nor in proportion to performance, but instead the sales, spoils or prizes go to the winner alone, the highest performing organisation on a particular occasion? Examples of such markets might be contracts for large-scale, one-off projects, such as building a new aircraft carrier or digging a tunnel under the English Channel. But other examples involve multiple events. In corporate recruitment the number of applicants may far outstrip the number of posts. Likewise, academic research councils regularly advertise funds for which the proposals received will outnumber the available grants. In such a market, if your organisation had the same distribution of performance as the other N organisations, it would share with them a chance of $1/N+1$ of coming first. What can be done to improve your organisation's chances?

At first sight, it would seem obvious: one should seek innovative routines that will increase one's *average*, or *expected*, performance. However, a simple simulation reveals that things are not quite so obvious if the new routines also affect *variance* in performance, that is, the expected deviation from mean performance. When choosing routines or a strategy in competition, one may sometimes be able to gain relative advantage by trading mean performance for variance in performance. Figure 4.20 shows four trade-off lines, for four different values of the number of competitors, N. The x-axis represents parameter values for performance *variance* and the y-axis values for performance *mean*. Different routines or strategies can be represented by different positions in this parameter space. Positions above and to the right of a trade-off line offer a chance of winning better than $1/(N + 1)$. Positions below a line offer a worse chance. The trade-off lines are estimated by simulating several thousand competitions. In each competition, performance values are generated for all $N + 1$ organisations. For N organisations these are generated by random sampling from a Normal distribution, with the parameters Mean $= 0$ and Variance $= 1$. For the remaining one organisation performance is sampled using variable Normal distribution parameters. After each competition it is recorded whether or not this special organisation achieved

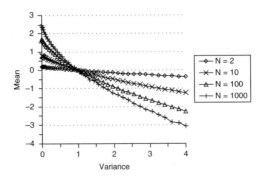

Figure 4.20   *In a winner-takes-all competition it can be advantageous to*
follow strategies that sacrifice some mean performance for
higher variance in performance. Trade-off lines in strategy
space are shown for various values of the number of competitors,
N + 1. Any player whose performance mean and variance
positions it above the trade-off line has a better than $1/(N+1)$
chance of coming first in the competition, given the other N
competitors follow strategies with mean = 0 and variance = 1.

the highest performance out of the $N + 1$. If multiple competitions are run
with the same parameter values for this special organisation, the chance
of coming first can then be estimated for that combination of performance mean and variance parameters. As can be seen from Figure 4.20,
given $N = 1000$ competitors, an organisation could choose routines that
raised mean from 0 to 2, and still suffer reduced chances of coming first if
at the same time the new routines lowered variance from 1 to 0. Reliable
performance (that is, performance with lower variability) can be reliably
inferior to its competitors' performances (that is, come first less often).
This is something for decision makers to bear in mind, given that stand-
ardising and simplifying routines, for example to improve the efficiency
and reduce the cost of their production, can reduce the variability in
routine outcomes. By contrast, if variance was increased from 1 to 2, any
reduction in mean less than about 1.5 would still leave the organisation
with improved chance of coming first. That is, it is possible for innovations to improve one's chances while lowering expected performance, if
at the same time they sufficiently raise variance in performance. Also, by
comparing the slopes of the trade-off lines it can be seen that the more
competitors one has, the higher the ratio between change in mean and
compensatory change in variance.

Recall from Chapter 2 the discussion, in the context of stochastic diffusion models, of the drunkard weaving from side to side along the middle

of a busy road: the function of an average is not the same as the average of a function (Savage et al., 2006). Just as there is an important difference between the drunkard's fate if he takes his mean path, and the drunkard's mean fate if he takes a random path, so we can distinguish between the competition outcome if we raise expected performance (the function of a mean), and the expected competition outcome of raising performance (the mean of a function).

Organisations whose managers have difficulties with the concept of this trade-off would be advised to avoid winner-take-all markets. Those remaining in them may need to support activities that are more variable in their effects on the organisation's performance. They may need to tolerate more experimentation, more diversity in approaches and ideas, and more mavericks proposing them.

CHOOSING BETWEEN LEVELS OF RISK

The question of whether organisations should engage in activities that are more uncertain in their payoffs deals with the concepts of risk taking and rationality. Biological organisms and species often find themselves in situations where the expected value of their current practices, i.e. the value under each chance event multiplied by the probability of that event, is negative. Innovations in behaviour can alter the expected value, but Slobodkin and Rapoport (1974) note that choices can be worth making even between actions identical in expected value. The key is to decide between more or less risky ones. Consider a gambler placing bets until he or she runs out of money. Each time a bet is placed, a fee must be paid (the ante), and the gambler starts with funds only a few times greater than this fee, and so can play only a few rounds unless they win one of them. The gambler is given a choice between a high-risk, high-rewards bet and a low-risk, low-rewards bet, both designed to be equal in their expected payoff. Despite their equality in expected payoff, the two bets do not lead to equally long games for the gambler.

Economists, following basic decision theory (Von Neumann and Morgenstern, 1947), might miss this point. Orthodox decision theory holds that rational agents should be indifferent between options identical in expected value. And so they should – in the long run. But what if there is not going to be a long run? Making any decision incurs a cost in itself, and when resources are particularly low, one may be unlucky enough to run out after just a few risks fail to come off.

Biological rationality, in contrast to economic rationality, dictates that taking small risks in the hope of winning small payoffs can be better than

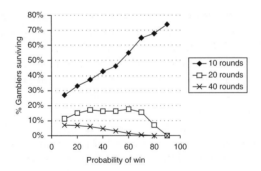

Figure 4.21   *Percentage of gamblers surviving after 10, 20 and 40 rounds*
of games with the same fixed expected payoff, ante and initial
funds, but a choice between various probabilities of winning
the payoff (x-axis). In this example, the ante is fixed at 4
and the expected payoff is 3, giving negative expected value
(-1) in playing a round. Given initial funds of 12, players
have the best survival rate from 10 rounds if they choose high-
win-probability, low-risk games, but games with more rounds
call for a lower-win-probability, higher-risk strategy.

taking large risks for large payoffs, since the small-risk actions leave you in the game for longer if you suffer a run of bad luck. What matters in biology is prolonging survival.

The point made by Slobodkin and Rapoport (1974) applies to cases where the expected value is positive or zero. Thorngate and Tavakoli (2005) use Monte Carlo simulation to extend the argument to situations of negative expected value. (On the website is our NetLogo version of their model, *LongRunGambling.nlogo*.) In such cases no one stays in the game very long, but those that last longest tend to have chosen riskier actions. So low is their initial resource level that only the big win that comes with a high-risk action can generate enough of a safety margin to survive for very long. The level of risk that yields the best chance of surviving the game increases as games get longer (Figure 4.21).

As Thorngate and Tavakoli suggest, this may explain why human psychology drives some poor people to take high-risk, high-stake gambles, such as robbing banks, compared with the activities of more affluent individuals. Those in an underclass may have negative expectations of life anyway, in which case live-fast–die-young risk taking can seem a rational strategy. Likewise, corporations with rather limited funds may need to revise their innovation strategy when the market outlook shifts from positive to negative. While a market is growing, new entrants should focus on low-risk small-scale sales targets – in contrast to the many Internet start-ups during the dot.com bubble of the late 1990s, who burned cash at crazy rates in pursuit of global market dominance. But when a corporation is in decline, desperate times may call for desperate measures, or high-stakes gambles. Such risk-taking behaviour may become increasingly likely if markets move towards a winner-take-all distribution of rewards, as may have happened in the banking industry before the financial crisis. For another example, if in the future only the top-20 universities are to receive research funding (the rest being reduced to providing mere teaching services), then most UK universities can expect to lose out. Those far outside the top 20 may have already accepted this situation and have opted to concentrate on teaching anyway. But others with ambitions to continue as research institutions may try to raise their ranking by spending money and incurring debts, such as by pushing out staff with poor research ratings, to be replaced by poaching highly regarded but expensive elite staff. In sport, association football clubs in England have become notorious for the debts they incur in bids to reach the top few places in the top league, and thus qualify for rich TV contracts and prestigious European competition places. At time of writing the efforts of several clubs have been rewarded with bankruptcy and the imposition of points penalties. Perhaps even the behaviour of dot.com companies becomes rational. Mindful of Brian Arthur's identification of positive feedback loops and market lock-ins in the IT industry (Arthur, 1989, 1994), they perceived the online market as a winner-take-all situation, and thus their expected position was to lose out. Only activities with high variance could hope to save them.

As organisations become more standardised and more reliable in their performance, their chances of survival in winner-take-all markets diminish. Future models of organisational learning need to pay more attention to the roles of chance in performance and selection by the market. Different market structures call for different attitudes to risk. Organisations need to consider their markets and their competitors' behaviour. They also need to reflect on their own sources of innovation and learning, and be prepared to adjust those sources in response to changes in the market.

SIMULATION MODELS AND REAL ORGANISATIONAL BEHAVIOUR

Building on the view of organisations as collections of individuals engaged in combinations of routines and heuristic searches for new combinations, computer simulation models of innovation within organisations have demonstrated how problem-solving and learning performance is sensitive to a variety of factors.

A metaphor with searching a landscape helped us to understand how a heuristic search method such as trial-and-error experimentation could lead to the discovery of satisfactory solutions in some contexts. But in other contexts it can leave one with an inferior solution relative to what others have discovered and with no path to further improvement. By learning from others, human workers are capable collectively of gaining knowledge to a degree not likely when they act in isolation, but they are then in danger of forming a consensus on a single solution before they have adequately explored the possibilities – the problem of premature convergence. The trick is to find a balance between, in March's (1991) phrase, exploration of new solutions and exploitation of existing ones.

Kauffman's (1993, 1995, 2000) *NK* fitness gave us a toy model of a search task involving seeking optimal combinations of interdependent parts, with the possibility of varying the numbers of interdependencies, and hence the difficulty of the task. There was also scope for varying the organisation of those interdependencies, to form tasks that might profitably be decomposed into subtasks, and solved separately from each other using methods that might not be feasible if employed on the whole task. Indeed, when agents are assigned different subtasks, the degree of interdependence between their subtasks could determine whether it was best for each agent to consider the fitness of the solution to their own subtask as their search goal ('egoism') or the total fitness of the solution to the collective task ('altruism').

Using a version of Kauffman's *NK* fitness, Lazer and Friedman's (2007) model showed the balance between exploration and exploitation was influenced by the social network structure between agents, including the network density and the way those links were distributed, the frequency with which agents used learning from others, the proportion of a solution they copied when they did imitate and the proportion of their own beliefs they mixed it with, and the preference for similar or dissimilar others when deciding whom to imitate.

The results from the Lazer–Friedman model sometimes showed considerable difference between the mean fitness of the organisation's agents, and the organisation's maximum fitness, the fitness of the best agent. This too varied with such factors as network structure, for example, a good maximum could be achieved if the organisation's network was split into fragments, but this would lead to a bad average fitness, since other agents were unable to learn from the agent lucky enough to have found an especially fit solution. The best approach to organisational learning for the organisation depended on whether the organisation needed one person with a fit combination of routines or many people.

Using another model from March (1991) we argued that organisations competing in winner-take-all markets might also need to deviate from the idea of pursuing the best mean fitness, in favour of routines with more variance in fitness, often less fit, but sometimes achieving the market best and thereby the rewards. With Thorngate and Tavakoli's simple model, we also noted the difference between economists' and biologists' concepts of rational decision making, and how, when expected returns are negative, the rational behaviour in the sense of prolonging survival may be to take big risks.

Summing up, then, those thinking about innovation in the form of organisational learning have to consider whether the organisation's employees are engaged in improving collective fitness, personal fitness, the fitness of the organisation's star employee (whoever that may turn out to be), the chance of being the organisation's star employee, or prolonging survival, whether of the organisation or the individual, in a winner-takes-all competitive environment.

Looking to the future, the lessons from experimenting with simulation models of organisations include that studying people's task structures and reward schemes will help us to understand their innovation and learning behaviour. In addition, studying their behaviour will give clues as to the task and rewarding structure in their market or environment. Empirical studies of people within real organisations, such as social network analysis of who communicates with whom and ethnographies of what activities people engage in, should address the question of what innovation behaviour is being shown in which organisations.

Of course, the model organisations discussed so far still make many simplifying assumptions. The organisation in the Lazer–Friedman model, for example, had a social network with fixed membership and learning behaviour, and its agents were homogeneous in their preferences for the various heuristic search methods and in their optimisation goal. But these are all assumptions that can be relaxed relatively easily in future models, to test sensitivity to them. In addition, sensitivity of results to the type of problem that agents had to solve, in Lazer and Friedman's model an *NK* fitness landscape, can be tested with variations on this landscape, including modular *SNKC* fitness representing decomposable tasks, as illustrated in this chapter. Other task definitions can be tried as well, including problems based on the satisfaction of constraints (see Chapter 6 and also Kauffman, 2000), and rugged landscapes based on fractal geometry (Winter, Cattani and Dorsch, 2007).

One contribution from ethnographic studies for modellers to consider is the studies of photocopier technicians at Xerox that reveal the extent to which problem solving in the workplace involve tacit knowledge and story telling (Brown and Duguid, 1991, 2000), rather than the beliefs encoded in official training and instruction manuals. Problem solving is a process of sense making, with solutions emerging from the interactions between multiple workers as they tell successive stories about what must be causing some problem, and try whatever actions each story implies should work. Problem solving is also entwined with social and cultural capital, as technicians draw upon each other's stories when constructing new narratives, and use the accounts of their own problem solving as conversational capital when socialising with other technicians. Modellers of organisational learning might like to consider how to represent problem solving when it takes the form of using metaphor and narrative.

Recent developments in organisational studies have also revisited the core notion of organisational routines (Cohen et al., 1996; D'Adderio, 2008; Feldman, 2000, 2003; Feldman and Pentland, 2003; Feldman and Rafaeli, 2002; Pentland and Feldman, 2005, 2008), in particular noting how innovation enters during the successive re-interpretation of routines, as people have to decide what counts as correct enactment of a particular routine, which they do under the influence of social relations and material environment. Just as Luc Steels's agent-based models of language evolution (Steels, 1996, 2002; Steels and Kaplan, 1998; Steels and McIntyre, 1998) built upon Wittgenstein's (Wittgenstein and Anscombe, 1953; Wittgenstein et al., 1978) remarks on rule-following and the use of symbols in both mathematics and everyday language, so future models of routine performance and learning in organisations may have to address what it is to perform the same routine twice, and who decides when a routine has been repeated and when there has been innovation. In Steels' model agents socially construct lexicons with enough shared meaning for useful communication to occur when problem solving. The words developed by these agents can be both synonymous and ambiguous in their meanings, and a certain amount of flexibility when applying linguistic expressions to currently observed situations may allow solutions from one problem environment to be reapplied by a process of analogy to newly seen environments, or even to fantastic or counter-factual ones. So, future challenges for modellers of innovation in organisational problem solving include the representation of narrative, reasoning by analogy, interpreting, performing and revising routines, and imagination.

5. Science models

Simulation models of science are an excellent basis for studying innovation processes. There are a number of reasons for this. First, science exemplifies innovation as a social and collaborative process. Not only are scientific advances the product of teams of researchers in laboratories, workshops, libraries and in the field, but also contributions build upon the past work of other scientists, 'standing on the shoulders of giants', as Newton once put it, but also on the shoulders of vast numbers of more modestly talented workers. Second, much scientific work leaves data trails, through which we can seek to trace its processes. The data includes the existence of institutions, such as industrial laboratories, universities and research centres, but the two most useful sources of data are academic publications and patents. Both publication and patent data can record multiple authors, indicating who collaborated with whom, and citations to previous publications or patents, indicating intellectual debts.

Within these data, patterns requiring some explanation may be discerned. These patterns include clustering among co-authors, academic papers and patents; geometric growth over time in scientific production; scale-free frequency distributions of papers per author and citations per paper; and various less familiar metrics from social network analysis for the author, paper and patent networks. How these patterns are generated and what this implies for science policy makers is a challenge that simulation models can help with. A simulation can demonstrate whether simple mechanisms are sufficient to generate these types of pattern, just as in Chapter 2 a variety of simulation models were invoked to explain bell-shaped and S-shaped curves. A model can provide sufficient explanation for some pattern without being the only sufficient explanation possible, and in that chapter we were left with several models with very different theoretical backgrounds but equal ability to fit the data. But with this caveat in mind, we may also note that knowing one sufficient causal mechanism may still be preferable to knowing none. So in this chapter we discuss models containing at least one generative mechanism for each of the above-mentioned patterns found in science.

If we focus solely on the patterns in the data, however, we risk neglecting important aspects of science, to do with its purpose. Like the organisations assumed by the models in the previous chapter, scientists are engaged in searches. There are problems to be solved, some of which matter to the world outside the science lab and the academic department. However, data on publications and patents do not include the difficulty involved in getting to the publication stage, nor how many resources it took. In particular, we do not know whether similar contributions to knowledge might have been made earlier or more cheaply, if the scientists involved had been organised in a different way. For some examples, could financial, human and time resources have been distributed better, and could scientists have benefitted from being exposed to different sources of ideas, via face-to-face conversations, lectures or reading? These what-if questions, similar to those raised for models of organisational learning, can be addressed by models of science as search activity.

In this chapter we describe simple generative mechanisms, before exploring the possibility of combining some of them into a single simulation model, one that both explains data patterns and offers suggestions as to the likely impact of organising science differently, thus becoming a tool for policy makers and their advisors. First, we present a mechanism for generating the scale-free frequency distributions found in academic publication data, based on a principle of the cumulative advantage, or the rich getting richer. We then describe a mechanism for generating clusters, based on the principle of homophily, or preference for interacting with others similar to oneself. To illustrate how organisation and resourcing can affect science as search, we reproduce models by Hegselmann and Krause (2006) and Weisberg and Muldoon (2009) in which different types of scientist agent, each with its own search behaviour, are mixed in various proportions to produce different search performance. One of us has already developed a model of academic science publication that generated both clustering and a frequency distribution (Gilbert, 1997). However, by combining more mechanisms, a much more sophisticated model of science can be developed (Watts and Gilbert, 2011), attempting to combine multiple examples of data fitting with heuristic search. We conclude the chapter with a review of our experience with this model, together with some references to other work on modelling science.

THE BIBLIOMETRIC TRACES OF ACADEMIC INNOVATIONS

When a scientist or academic researcher innovates – when he or she discovers something new or creates some novelty – describing it in a book, journal paper or conference proceedings alerts the rest of the world to the innovation. This affords the opportunity for other workers in the field to verify the innovator's claims, to give credit to the innovator for being first to make the discovery or for attaining a good standard of work, and to incorporate the novelty in their own research and teaching, where it may lead to further innovation. If innovative work is rewarded with new resources, then making one discovery may lead on to the funding of future ones. Commercially useful innovations may appear first in patents, or be kept secret within the corporation that hopes to turn them into products and services. But much innovation produces publications, and those wishing to understand how innovation is generated, diffused or made use of may seek answers in bibliometric data, or records of publications.

Today such information can be found in online electronic databases, including *ISI Web of Science* and *SCOPUS*. But attempts at measuring science – or scientometrics – pre-date these. Price (1983) noticed from the heights of stacks of journal volumes that the number of pages written in these volumes was increasing geometrically, or approximately exponentially. That is, the size in pages of each year's volume was a constant multiple of the size for the previous year. Armed with a tape measure, further investigations at the library revealed that this phenomenon was common to different journals, in various scientific fields, and that geometric growth was also occurring in the numbers of papers and the numbers of authors (Price, 1963). It is as if each page of science (or paper or author) has the effect of inspiring some number of further pages (or papers or authors), and the number generated is greater than one.

Earlier than Price's discoveries, Lotka (1926) examined catalogues of publications in chemistry. He documented that the numbers of papers per author seemed to follow *an inverse power law distribution*. That is, if x is the number of papers per author, then y, the number of authors with x papers to their name, is given by:

$$y = A. x^{-\beta}$$

Figure 5.1 shows one of his datasets with a power law fitted to it using Excel. For the exponent, the parameter β is approximately 1.8. Note how this tends towards a straight line when both axes take logarithmic scales (a 'log–log plot').

Power law distributions have some unusual mathematical properties. When sampling from some population that is power-law distributed, the sample mean, or average, increases with the size of the sample. In contrast to the distributions known from high school mathematics, such as the *normal*, *poisson* and *exponential* distributions, power law distributions lack a characteristic scale, and are often referred to as *scale-free distributions*.

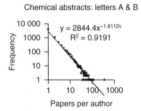

Figure 5.1   Lotka's law: the frequency distribution of the number of papers per author follows an inverse power law.

They occur in a variety of natural and social phenomena, including the occurrence of words, the number of hyperlinks to webpages and the sizes of cities and earthquakes (Barabási, 2002). A familiar implication of a power law is Pareto's (1909 [1972]) observation that 20 per cent of the population owned approximately 80 per cent of total wealth, in the case of land: the so-called 80:20 rule. Power-law distributions indicate inequality; a small minority is responsible for a large majority of cases. Price (1976) identified power law distributions in the numbers of citations received by papers. Most papers received few citations, if they were cited at all. The bulk of the citations are made to a small number of papers – what have been called 'citation classics'.

Along with growth curves and frequency distributions, bibliometric data also supply us with information about networks. As Price (1965) noted, citation relations – the relations formed when one paper refers to or cites an earlier one – constitute a network. Figure 5.2 illustrates this and some of the other network relations obtainable from bibliometric data.

The picture focuses on just two authors and six papers. Agent A1 is the author of three papers, P6, P4 and P5. Authorship relations are shown with thick lines with arrowheads. Agent A2 is also an author of P5, so A1 and A2 are linked by a co-authorship relation (thick dashed line). (The authors of P1, P2 and P3 are not depicted.) Paper P4 refers to papers P1 and P2. Reference relations are shown with thin lines with arrowheads. P5 also refers to P2, so P4 and P5 are linked with a co-citation relation (thin dashed line). P6 is linked with a co-citation relation to P4, since they both refer to, or cite, P1. Co-authorship and co-citation are undirected relations: A1 co-authors with A2 if and only if A2 co-authors with A1. So their lines have no arrowheads. Other relations could include that between two papers which share an author, and that between two papers that are both cited by the same paper (called 'bibliographic coupling'), but these relations are less frequently analysed. In addition to relations between papers and authors, when bibliometric data also include keywords, or

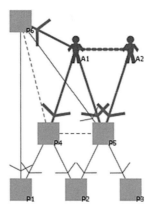

Figure 5.2 Bibliometric nodes and links. Picture shows two authors (shown as 'person' shapes'), six papers (squares), and authorship, reference, co-authorship and co-citation relations (thick arrows, thin arrows, thick dashed lines and thin dashed lines, respectively).

titles, abstracts and other text from which keywords can be identified, then papers can be related by their contents, and authors by the topics they write about. Journal papers can also list addresses and institutions for their authors, from which geographic relations could be derived, but changes of university are quite common among academics, and the institution listed at the time of publication may not be the one where the author conducted the work described.

If we have data on who wrote what, who cited what, and possibly also on what they wrote about and where they are located, what can we infer about innovation? Papers serve different purposes, of course, but common functions include introducing innovations, and reviewing previous innovations. Of the first type, papers may announce a recent discovery or observation, introduce a new concept or inference from existing theory, make a conjecture that calls for further discussion or investigation, or identify some previously overlooked flaw in past work. Of the second type, we find literature reviews, summarising, often critically, the work of others, often with a view to inspiring future innovations based on the reviewed work. In both cases, the aim may be said to be to make an original contribution to knowledge, though the work described may have already been introduced to other audiences in other papers, and is only expected to be original to the intended readership of the new paper. If one can identify what the innovation is in a particular paper, however, then papers that cite either that innovation or that paper represent the diffusion of the innovation.

Co-citation relations linking pairs of papers which cite the same paper may indeed indicate that the citing papers share a topic or idea. But there may be more than one function for the references in papers. Robert K. Merton (1968a, 1988) treated references as the giving of credit where it was due. Gilbert (1977) suggested they had a rhetorical function: the citing author is claiming membership in a social clique for him or herself, or perhaps for the cited author. Of course, these two functions may coincide.

Referring to a paper does not guarantee that one has actually read the paper, let alone understood its authors' messages, though we tend casually to assume that that is what is implied by the reference. Some authors may be copying references from other authors' papers, assuming that something well-cited by them is too important to be left out and must have been crucial to their argument. But without studying the context of the reference, that is without reading what a citing author thinks of the referenced item, one cannot tell from the mere fact of referencing whether the citing author is endorsing and building upon the cited work, or criticising it and defining their own position in opposition to it.

The growing amount of work in the analysis of bibliometric data indicates there remains much hope that academic papers represent innovations, and that data on the occurrence of papers tell us something about the production of innovation, including its rate and its sources. If experiments with computer simulations of scientific production could demonstrate some of the processes that will or will not generate artificial bibliometric data with patterns similar to those seen in real data, insight might be gained into the extent to which publications represent useful problem solving, social clique formation, and the rewarding of past success.

What micro-level processes or mechanisms could generate the macro-level phenomena? In the following sections we introduce some algorithms for mechanisms that can generate something similar to what appears in the data. In each case we suggest behavioural practices that could form such mechanisms in the real world. These are not intended to be the only practices that could generate the patterns, nor need the algorithms be the only generative mechanisms. When alternatives have been suggested they can be compared to see which is more plausible. Until then, we shall treat a possible explanation as being better than no explanation.

We illustrate the mechanisms with data from a single journal, downloaded from the *ISI Web of Science*. The journal, *Research Policy*, is one of the core journals serving the field of innovation studies, and from its foundation in 1974 has grown each year. After 30 years it had reached about 1400 papers by about 1500 authors, a scale well within the capabilities of the simulation model we shall present later in this chapter.

CUMULATIVE ADVANTAGE

The generative mechanism for scale-free distributions presented in this section is the result of several developments. Herbert Simon (1955b) was the first to propose a mechanism for generating a power-law distribution, in this case Lotka's (1926) frequency distribution of papers per author. When applying it to citations per paper, Price (1976) described it as a process of *cumulative advantage*. The rich get richer. New citations tend to go to those papers already rich in citations. Likewise, opportunities to publish papers tend to go those authors already prolific in publications. Such cumulative-advantage processes had already been identified in science by Robert K. Merton (1968, 1988), who called it the *Matthew Effect* after the lines from the Gospel according to Matthew:

> For unto every one that hath shall be given, and he shall have abundance: but from him that hath not shall be taken away even that which he hath.' (Bible, King James Version, 25:29)

Gilbert (1997) reused Simon's algorithm in a simulation model of the structure of academic science, the GASS model, which will be mentioned in the next section for its additional modelling of clusters. Barabási and Albert popularised a version of the mechanism in the context of network evolution (Barabási, 2002; Barabási and Albert, 1999). They described the process as one of *preferential attachment*: a network is grown by linking each new node to one selected with preference for the number of links it already has. This has led to a variety of network models with scale-free distributions of links per node, most of them, but not all, using preference for links as part of their generation (Caldarelli, 2007; Dorogovtsev and Mendes, 2003), although, as discussed in Chapter 3, none of them provide a realistic representation of a social network. A problem with these models is that they assume that nodes know how many links all the other nodes already have, so that they can find those with many links. Pujol et al. (2005) extended this network modelling to social interactions based on the prisoner's dilemma from game theory. Their model has the advantage of not assuming that agents, or nodes, have knowledge of the numbers of links of the entire network, an assumption that for a network of thousands of authors would seem implausible. Bentley et al. (2011) also avoid this assumption. They do this by representing the process of adding new nodes as one of *imitation* of previous cases of node addition. One needs only to sample a previously added node and find out which node it was linked to. Knowledge of how many other links that node had is not required. A past node addition to be imitated could be sampled from all such past additions, but Bentley et al. (2011) add a plausible restriction to the algorithm. Only node additions made within the past m iterations can be imitated by the current node addition, where m is called a memory parameter. There are other ways to bias this sampling towards recent cases. The model of Watts and Gilbert (2011) follows the science model of Boerner et al. (2004) in using a Weibull-shaped ageing function to weight the past cases. This takes two parameters, alpha and beta, and, depending the choice of parameter values, is capable of emulating a negative exponential function (when beta $= 1$) or a bell curve. Usually it resembles a bell curve with one tail stretched out, representing in this case a higher chance than normal of particularly old papers being selected. However, the simpler mechanism of Bentley et al. (2011), can still, depending on its parameters, generate a wide range of frequency distributions.

The mechanism is a stochastic process – that is, it involves random processes – and stochastic processes are usually introduced in terms of coloured balls being drawn from and added to urns. For our purposes here, we assume we are adding papers to authors, though adding references to papers might employ a similar process.

The NetLogo program *PapersPerAuthor.nlogo* on the website demonstrates the algorithm. Figure 5.3 and Figure 5.4 show example output fitted to the first 30-years worth of data from a real journal, *Research Policy*. At initialisation, a number of urns are created. Call them the authors of *foundational papers*. Thereafter we simulate a number of iterations, which represent time steps such as years. At each iteration, a batch of new items (papers) is added. The size of a batch grows over time to represent the geometric growth observed by Price. For each added paper in a batch we need to decide which urn (author) it will be added to, that is, which author has received the opportunity to publish a paper. With some given chance, we create a new urn and add the item to it. This represents a new author

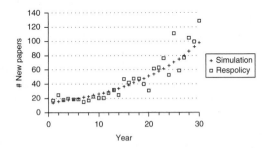

Figure 5.3   The number of new papers published by year in the journal
Research Policy ('ResPolicy') and generated in the simulation
PapersPerAuthor.nlogo. *Both show geometric growth over time.*

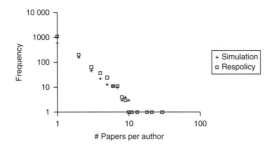

Figure 5.4 *The frequency distributions for numbers of papers per author in the journal Research Policy and in the simulation. Both resemble a scale-free distribution. Note that the axes have logarithmic scales.*

entering the field with his or her first publication. If the author is not new, however, we have to select an existing one. We do this by sampling a *recent* paper and copying the author from it. In pseudo-code the algorithm for generating a power-law frequency distribution can be described as follows:

```
; Papers per author simulation
Initialisation:
        Create founding papers
        For each founding paper
                Create one author for that paper
        Next founding paper
Simulation iterations:
        For each time step
                Create a batch of papers
                For each paper in batch
                        Add an author to paper:
                                If chance < some given probability
                                then
                                        Return new author
                                Else
                                        Sample a paper from a pre-
                                        vious time step, weighting
                                        the papers by the recency
                                        function.
                                        Return the author from the
                                        selected paper.
                                End if
```

                    Next paper in batch
                    Increase paper batch size according to
                    field-growth.
            Next time step
    Final:
            Output number of papers per author
    End.

The *recency* function for weighting past papers is based on the age of
the paper, that is, the number of time steps since it was created. For this
function we can use a simple cut-off, as Bentley et al. (2011) did: if age-
of-paper $<=$ m then weight $= 1$, else weight $= 0$. As indicated above,
Boerner et al. (2004) and Watts and Gilbert (2011) instead employ a
Weibull-shaped function of the age.

As well as the *recency* function parameters, the algorithm needs the
number of founding papers, the number of time steps, the number of
new papers in a batch and the rate at which this grows over time, and
the chance of a paper's author being new to the field. A relatively simple
enhancement is to create papers with multiple authors.

How might the model's processes be interpreted? Letting the field
grow over time implies that on average one publication causes more than
one later publication, perhaps as criticism of the first paper, elaboration of its points, or simply because following the first publication the
field was rewarded with the resources to enable further research to be
conducted and papers produced. The usefulness of a *memory* parameter
or a *recency* weighting function indicates that causal relations between
publishing events have a time scale. Linking authorship of one paper to
authorship of a past paper implies that the distribution of resources for
paper production and publication is influenced by past production. This
is the cumulative-advantage process for authorship, the Matthew Effect.
Perhaps the knowledge obtained while producing the first paper enables
an author to discover further contributions. Perhaps reputation gained
from the first publication leads to the author attracting funds and talented
co-workers to enable further research. Perhaps improved reputation for
the author makes it easier to persuade peer reviewers to trust further contributions from this author, though by removing the author's name from
a paper submitted for review, journals can attempt to reduce this source
of influence. However, the popularity of a previous publication can still
raise the chance of acceptance for further work in the same research area.

The simulation *PapersPerAuthor.nlogo* is able to generate data in which
a few authors are highly prolific, while most publish very little – just like
real bibliometric data. Such a mechanism can also generate data for the

number of citations per paper following a realistic scale-free frequency distribution (Watts and Gilbert, 2011). But no discriminations in quality between authors or papers have been made. Differences in talent are not required to generate the scale-free distributions seen in for papers per author and citations per paper. Just because some author is highly prolific and cited in some scientific field, we cannot infer anything about his or her talents, quality of work, or efficiency of organisation. This does not imply that such a person is not more talented than others; only that a scale-free distribution does not need this for an explanation.

This should be born in mind by policy makers considering distributing research funds on the basis of past performance judged by papers or citations. It is not so much that the best innovators are rewarded for their past successes as that the rewarded become the future innovators. By choosing to reward researchers for past publications, and restrict resources from the less prolific researchers, policy makers can make it a *self-fulfilling prophesy* (to use another phrase from Merton) that previously successful innovators will be the ones most likely to succeed at innovation in the future, and a self-fulfilling prophesy that those whose searches have yet to yield solutions worth publishing will not have the resources to seek and find solutions in future.

CLUSTERING AND GROUPS AMONG SCIENTISTS

The simulations that generated scale-free frequency distributions of publication per author and citations per paper hinted at processes that allocate resources and attention in science on the simple basis of past success. Evidence of geographical, institutional and social clusters in scientific production points to further determinants of who, where and how often publication occurs and citations are received. In this section we examine models of cluster formation.

Academic scientists, like academic researchers in general, are clustered in many ways, including socially, institutionally and by the topics they study. Institutionally, they may be organised into universities, with buildings geographically clustered. Within a university such as that of the authors, divisions are made to form faculties ('Arts and Human Sciences', 'Engineering and Physical Sciences', 'Health and Medical Sciences' . . .), and then departments ('Economics', 'Sociology', 'English', 'History', etc.) These divisions may be made to make management of the academic staff more practical, and they are often reflected in the geographical location of the staff – the buildings they share or neighbour – but the divisions express some opinions concerning the contents of research and teaching performed by the staff, and how these may be affected by geographical location. For example, by placing Sociology in the same building as Economics and Psychology we make it easier for their respective staff to encounter each other, attend each other's seminars, and share students. A few universities in the UK have a rival division into colleges, most notably in Oxford and Cambridge where each college has its own set of buildings and thus its own opportunities for socialising for teachers and researchers in multiple academic subjects. In the case of the University of London, the division into colleges has become so much stronger than that by subject matter, that they now appear in league tables of universities as institutions in their own right ('Imperial College', 'University College London', 'the London School of Economics', etc.).

The evidence for such organisation of academics is easy to come by – the universities' websites provide details of their departments and staff lists. An alternative organisation of their teaching and research may be found in their libraries, in the decisions concerning where to locate books (usually based on some standard classification system), how much space to allocate to particular classes and how much to spend on new additions. Again physical location reflects some thinking on which subjects should appear close to which, and increases the probability of potentially productive encounters, such as when a researcher in one field chances upon a book in another field (but on a bookshelf he or she has had to pass) that offers them some contribution – an analogy or metaphor to apply, or an unfamiliar model or technique. If innovations come from novel combinations, a multi-disciplinary space such as a college common room, a library or a university bookshop may have an important impact on innovation production, especially when increasing use is made of electronic databases and encyclopaedia, and search engines that can take one more directly to one's official subject area, without the same opportunities for chancing on something else.

Electronic data sources, however, have led to an explosion of interest in studying scientists' organisation. When scientists publish papers with other scientists, and when they cite works by others, they create social relations the details of which we may record. The tools of social network analysis (Carrington and Scott, 2011; Wasserman and Faust, 1994) can be applied to data on these relations, with a view to understanding how network properties contribute to the productivity of the scientists. The studies of authorship and citation networks in science (Price, 1963, 1965, 1976; Small, 1973) pre-date electronic databases, and Collins (1998) needed no electronic sources to trace networks of master–pupil relations among philosophers across national borders and back to the time of the Ancient Greeks. But with databases like the *ISI Web of Science* and *SCOPUS* any researcher with access can download datasets relating publications to each other and resulting from searches based on particular authors, journals, keywords, classifications or years.

Displayed visually the network data become 'maps of science' (Boerner et al., 2012; Shiffrin and Boerner, 2004; Wagner et al., 2011), indicating how near or far apart disciplines are, in so far as one cluster of papers or journals can be distinguished from another. Universities' organisation already gave one indication of divisions based on the contents of science – the names of departments and research groups indicating fields and sub-fields. But the existence of inter-departmental research seminars or centres as joint ventures that cut across these disciplinary boundaries indicates that the constructions that are the distinctions between academic disciplines may be artificial. The citation networks offer an alternative view on who actually works with whom, whether inter- or intra-department and discipline, and whose work depends on who's for the justification of its arguments. At present the datasets have problems – for example, in trying to match 'Smith A' to 'Smith, Andrew' you risk mis-identifying authors, especially if more than one person called A. Smith publishes in a particular field. The researcher faces a choice between, on the one hand, working on a subset that is small enough to be cleaned and verified in a reasonable length of time, and on the other hand, working with a larger dataset, perhaps running to millions of records, and a few computerised processes for cleaning and transforming them automatically, hoping that any biases or faults in the processes may be offset by its sheer size. Either way, studies so far show that scientists can be clustered by the contents of their publications, their citations and their co-authorship relations. These network properties can also be correlated with indicators of academic performance, such as the number of papers produced by an academic (a measure of quantity of research), and the number of citations they generate in other authors' works (perhaps a measure of quality, although see Fuller, 2000, p. 5ff, for objections). This invites the question whether initiatives that alter the networking behaviour of academics – such as reorganising departments' locations, funding trips to meet face-to-face with potential collaborators, inviting prominent and prolific experts to visit, and even reading the results of bibliometric analyses of one's field – might improve knowledge contributions and their influence.

At time of writing, however, while network analyses can identify patterns, we do not yet know what those patterns mean. A correlation between network properties and scientific output is not the same as a causal link. Just how there come to be the patterns in the network and in the academic production is not yet understood. This is where simulation models can contribute. We can take hypotheses about the mechanisms generating these patterns, represent them in computer code, and attempt to produce under the controlled conditions of the simulation some patterns analogous to the ones observed in the real-world data. Some of the patterns and the encoded processes may also have analogues in other domains than that of scientists' networks. Clustering has been related to innovation in economics and geography, for example by Porter (1990) and Krugman (1991).

How is clustering among researchers produced? One approach to explaining the existence of clusters is to invoke the existence of something external to the researchers. For example, researchers are people with the usual social and economic needs. They need space for their buildings, homes for their families, roads to travel along between these, and non-academics to supply them with goods and services, all of which cost money. Hence, geography and economic factors influence where researchers can maintain their work and at what cost in human and financial resources. Most, though not all, universities are located in towns and cities. The top research universities in the world are not distributed evenly across the globe, but instead cluster, for example in New England in the United States. Analysis of university league tables or rankings reveals a disproportionate number of them to be in English-speaking countries. The *Times Higher Education World University Rankings* top-20 lists for 2012/13, 2011/12 and 2010/11 contained only one university from a non-English-speaking country, ETH Zurich, located in multi-lingual Switzerland (http://www.timeshighereducation.co.uk/world-university-rankings/). So language seems to play a significant role in determining who is perceived as successful.

One further influence on researchers may be the world they study, that is, the subject matter of their research. By this argument, the researchers successfully publishing are those who happened to search in the right place. They used the right techniques to solve problems, or they attempted the problems that were most ripe for solving. The structure and distribution of researchers and their publications reflects the structure of an underlying reality. Reality is divided into the physical, the chemical and the biological, and so researchers tend to be physicists, chemists, biologists, etc. But before making statements about how reality is structured, we should examine the extent to which processes within the research community itself can explain the disciplinary patterns that are formed.

Another type of approach to explaining clusters, then, is to invoke something internal, or endogenous, to research production. Social network theory suggests a number of processes which might be transferred to the context of scientific production. They will be referred to here as learning/influence, structural balance and homophily.

The explanation for clustering based on learning or influence holds that when seeking a new person to co-author with, a new topic to research or a new paper to respond to, we are influenced by those we are already linked to. Social influence processes were discussed as part of epidemic diffusion models in Chapters 2 and 3.

A second theory about clustering in social networks is the theory of structural balance. This is primarily a theory about what types of three-person clusters, or triads, will tend to occur in a network. Consider three researchers, Adam, Betty and Craig. Adam feels positively about Betty. Betty feels positively about Adam. Betty feels positively about the third person, Craig. If Adam and Craig feel negatively about each other, then that is going cause tension between Adam's feelings about Betty and his feelings about Craig. If one of these two sources of feelings were to change, the tension would be resolved. For this reason, triads that are made up of two positive relations and one negative relation are expected to be rare in real social networks. It would be surprising if Adam endorsed Betty's work, and Betty endorsed Craig's work, but Craig criticised Adam's. By contrast, two negative relations might raise the chance of the third relation being positive, by a principle of 'my enemy's enemy is my friend'. Two critics of the same author or paper have something in common with each other. Analyses of networks obtained from bibliometric data, paying particular attention to the frequencies with which particular types of triad occur, might give us some indication of whether structural balance processes are influencing co-authorship and citation. Identifying which citations are positive and which negative, however, requires some understanding of the contents of the papers making the references, and currently it is hard to perform this contents-analytical task on a large scale with a reasonable degree of accuracy. Co-authorship is also ambiguous in whether it is positive or negative. Sometimes intellectual opponents are invited to debate their differences in a single paper.

With this in mind we turn now to the third explanation for clustering, *homophily*, or the preference for similarity. This concept has been described by several sociologists (Homans, 1951; Lazarsfeld and Merton, 1954; McPherson et al., 2001; Tarde, 1899). Rogers employed homophily thus: 'the exchange of ideas occurs most frequently between individuals who are alike ... Such similarity may be in certain attributes, such as beliefs, education, socioeconomic status, and the like' (Rogers, 2003, p. 305). Various agent-based models have shown that combining a process of imitation or influence with a process of seeking interactions from those similar to oneself is sufficient to lead to the emergence of clusters of relatively similar agents. In this section two classic models will be described.

Axelrod's model of cultural influence (Axelrod, 1997a, Chapter 7;

1997b) is an agent-based model based around this principle (see *ACM_ Similarity.nlogo* on the website for our version). In it, a population of agents, initially heterogeneous in their cultural traits, engage in pair-wise social interactions. During an interaction, the two participants compare their traits in one randomly chosen cultural dimension or feature. If they hold the same trait in that dimension, the interaction continues to a second stage. In this, one of the participants (it does not matter which) attempts to imitate the other. That is, the imitating agent copies a trait from the imitated in one randomly chosen dimension where they differ in traits, if any such dimension exists. So the first stage of the interaction involves a test for similarity – the more dimensions they share traits in, the more likely it is the interaction will continue. The second stage involves imitation, which has the effect of making the two more similar than before. There is a positive feedback loop here: similarity causes interactions; interactions cause similarity. Axelrod found circumstances in which, from the initially heterogeneous population, cultural groups of agents emerged, which he called 'regions'. Within the group all members shared traits in all dimensions. Members in different groups lacked shared traits in any dimension, and thus interactions now had zero chance of passing the similarity test.

Later investigations by others revealed that the most likely number of groups to emerge from a population of a given size was determined by the ratio between two parameters, the number of traits per cultural dimension or feature, and the number of features (Castellano, Marsili and Vespignani, 2000). When this ratio is low (for example, 2 traits per feature/2 features = 1), the agents form a single homogeneous group. When it is high, (e.g. 256 traits/2 features = 128) the agents might fail to find a single other agent to interact with, and a population of 100 agents would represent 100 groups. Between these two extremes, the behaviour of the model underwent a phase transition, with the number of groups increasing in an S-shaped curve as the traits: features ratio increases. The exact shape of the curves and the take-off point could vary with the network architecture between agents (e.g. two-dimensional four-neighbour grid, one-dimensional ring, complete network), but the presence of the phase transition remained clear (Klemm et al., 2003). Another phase transition can be traced if one alters the similarity test. As described above, 'similar agents' continued to the imitation phase of an interaction if an initial comparison of the cultural traits in one feature resulted in a match. If agents have more than two cultural features, then we can compare traits in more features, and raise the threshold at which we say two agents are sufficiently similar to more than one matching feature. Figure 5.5 shows how increasing the size of this *similarity threshold* moves from populations tending to form a single homogenous group to their remaining stuck with

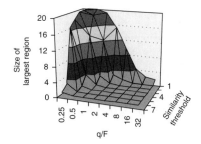

Figure 5.5   *In Axelrod's model of cultural influence, the size of the largest cultural group depends both on the ratio of traits (q) to features (F) and on the threshold similarity, shown here as the number of matching features required for imitation or influence to occur during interactions.*

many one- or two-agent groups unable to establish sufficient commonality to be able to communicate. Hence a number of options exist in Axelrod's cultural model for affecting the number of emergent groups.

Opinion dynamics models can also relate group formation to a preference for similarity. Whereas Axelrod's model represented agents' cultural beliefs as discrete values or traits, the agents in opinion dynamics models have opinions that are represented as continuous variables with values in the range 0 to 1. These could be interpreted as ranging from 0 = 'Extreme left-wing' to 1 = 'Extreme right-wing', or from 0 = 'Strongly disagree' to 1 = 'Strongly agree'. In the simplest examples of opinion dynamics models it is assumed the agents have opinions in just one dimension only. This avoids the tricky question of how opinions in different dimensions might interact with each other, and also enables us to display evolution in opinions on a two-dimensional chart, with time (simulation iterations) along the x-axis. Numerous examples of these simple models exist, including some with analytic results as well as simulations, but we shall concentrate here on one by Hegselmann and Krause (2002, 2006). The website contains our versions for Excel/VBA and NetLogo (*HK_OpDynBC.xls* and *HK_OpDynBC.nlogo*).

At the start of the simulation, agents are given opinion values drawn from a uniform distribution. At each time step, each agent examines the opinions from the previous time step and adopts the consensus view – the mean value – of all neighbouring agents. A neighbourhood is defined by 'confidence bounds': an agent has confidence in other agents only if the opinions of the others are located no more than some given amount, the confidence bound, from its own. Opinions outside this bound are considered too remote to be understood and their holders too different to be trustworthy, and so are ignored.

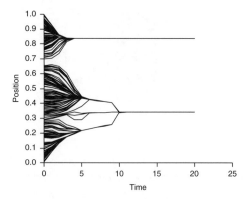

*Figure 5.6   As time goes by, a population initially heterogeneous in its
opinions or positions forms a small number of groups with
consensus positions. In this example, two groups have emerged.*

Figure 5.6 shows an example simulation run, with 255 agents scattered
across the opinion space at the start (left-hand side), and then converging
on two eventual consensus positions by the end of the run (right-hand
side). Judging from the number of lines flowing from the left, the lower
consensus is held by approximately 60 per cent of the population in this
example. The agents at one consensus position are now outside the confidence bound of those at the other position, and so can exert no further
influence on them. Hence the population has become polarised and frozen.

How many consensus positions one may expect to obtain depends on
the size of confidence bound held by the agents – the two seem inversely
proportional for most values, though there is relatively high variability for
bound sizes between 0.2 and 0.3 (Figure 5.7). With small bounds, there is
more chance that other agents escape your confidence bounds, leaving you
in smaller groups and pulling you less distance from your initial opinion.
With larger confidence bounds, there are more agents you are willing to
listen to, and consensus groups end up much larger.

In this model of opinion dynamics, the influence of agents with a particular opinion is determined by population density – if there are more agents
within one's confidence bound to the left than to the right of oneself, then
one will be pulled towards the left. Some of those agents to the left may
themselves be being pulled left, and in later iterations may be pulled out of
one's confidence interval. But while the majority of influences remain on
one's left, that is where one moves. Maintaining one's opinion is hard – it
requires a balance between the neighbours on the left and the right, and on
that balance persisting over time.

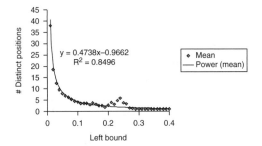

Figure 5.7   The number of distinct consensus positions emerging from the opinion dynamics declines as the confidence bound defining similarity gets wider.

The opinion dynamics model represents beliefs as positions in space. Another model invoking homophily and representing beliefs by their spatial location is Gilbert's Academic Science Structure (GASS) model (Gilbert, 1997), which employed x and y dimensions to represent the contents of academic papers as points in a knowledge space. The website contains replications of this model in both *NetLogo 3D* (*gass.1 3D.nlogo*) and Excel/VBA (*GASS.xls*).

The simulation is based on the idea of the construction of a new paper being inspired by an existing paper, the 'generator paper'. The simulation's mechanism is based on Simon's algorithm (Simon, 1955b) that explains Lotka's Law concerning the power-law distribution of papers per author. It is similar to the algorithm described earlier and used in *PapersPerAuthor.nlogo*, but each paper has only a single author, and each author is given an 'expiry date', after which he or she will have left the field, for whatever reason.

To produce clusters of papers, analogous to subfields, further principles are introduced. In determining a new paper's contents (x and y coordinates) the contents of the generator paper (and its author) are taken as a starting position. But the contents are then modified under the influence of a number of other papers – its references. These reference papers must be sufficiently relevant in contents to the new paper, however. Relevance is represented in the model by distance. So reference papers are chosen probabilistically from all papers within a given radius of the starting position. The influence of the referenced papers is important, as without them the new paper would still have the position of the generator paper. To be accepted for publication it needs to be judged to be sufficiently original in content, and so new papers are cancelled if their position is not more than a given distance from all other papers' positions. If an area of the field

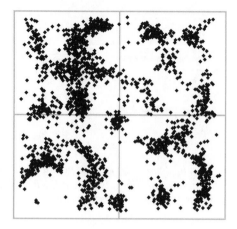

*Figure 5.8   A simulation of a scientific field from the GASS model, with
papers forming clusters or subfields.*

becomes too densely packed with papers, finding an original position may
become too difficult and publishing in this subfield will die out. Thus a
new paper's position is determined by the generator paper's position, the
influence of the references and the requirement for originality. Figure 5.8
shows the resulting clustering of paper points from a typical simulation
run of the Excel version. The program itself allows the publication dates
of papers to be displayed as position in a third dimension.

Clearly, then, a tendency for authors or papers to resemble earlier ones
can lead to the appearance of clustering of papers. It is not necessary to
invoke any sources of influence over the contents of papers external to
the people and processes that produced the papers, though these may still
exist.

But models based on the principle of homophily leave us with two new
puzzles: minority influence and the diffusion of innovations across cultural
boundaries. In the opinion dynamics model there was no scope for what
social psychologists sometimes call 'innovation', or 'minority influence',
when a group exists in which a majority share one opinion, but are converted to an opinion held initially by a minority (Dreu and Vries, 2001). If
we want to model the diffusion of innovations such as new ideas, we will
need a model in which agents' influence over each other depends on more
than just how many hold each opinion and how different those opinions
are from each other. The Axelrod model also leaves us with a puzzle: once
cultural groups have formed, how can there be any further change? If an
agent in one group develops, by chance mutation, a novel idea or trait,
how can this diffuse across the cultural boundaries between groups? While

boundary spanners remain, with traits in common with members of multiple groups, these agents can still interact with and transfer traits between groups. But in the Axelrod model such agents disappear, leaving the boundaries fixed. Klemm et al. (2003) added a trait mutation process to the Axelrod model. At a high mutation rate, this prevented the formation of groups with hard, well-defined boundaries between them. Meanwhile at a low mutation rate, mutation could recreate boundary spanners with the potential to dissolve boundaries, allowing everyone eventually to form one single homogeneous group. However, the most likely outcome of a mutation event was that the agent introducing the novelty promptly loses it again through imitating its neighbours. There is no sense here of the innovation being a good idea, or a superior solution to life's problems.

One suggestion from modellers is that minority influence can follow if the minority are more certain or persistent in the expression of their views, or less sensitive to the views of others (Deffuant et al., 2002; Weisbuch, Deffuant and Amblard, 2005; Xie et al., 2011). Models in this vein may try to explain the spread of ideas from extremists, such as terrorists and religious fanatics.

Another approach may be to consider the role played within groups by hierarchy, power and conflict. Those in positions of power, perceived or actual, may be able to introduce an innovation and persuade those people below them in a hierarchy to endorse it. If the homophily principle is operating, it might be expected that successful introduction of innovations tends to be made by those perceived as the most central to the group, its leaders who best represent the group's identity and norms. More peripheral members trying to introduce a novelty may be turned on by the other members as deviants corrupting the group. Exercise of power over others may cause resentment, especially if they have previously experienced power themselves, either in this group or another. If one such resentful member challenges a group leader, other members may determine, by their choices of whom to endorse, whether the usurper is rounded on and punished, the group leadership changes, or the group splits in two, with the rival leaders each taking some of the members with them, who thereafter have a stake in endorsing the opinion of their chosen leader. Contrast this picture with the dynamics of traditional epidemic-style diffusion models, in which a novel idea diffuses through a population one node at a time. A richer source of group dynamics could lead to more dramatic divisions between adopters and non-adopters. Also contrast it with models that represent innovation by using random mutation events, where again adopters appear gradually, and scattered across space and social networks. Conflict, between groups or between individuals, energises, and is invoked as a major source of intellectual production in Collins's sociology of philosophies (Collins, 1998).

Along with the puzzle of minority influence, the models in this section raise the question how, once agents have converged on a few cultural positions, ideas can diffuse from one group to another. In Axelrod's cultural model, for example, the boundaries between cultural groups became fixed and imitation or influence ceased. There is some empirical evidence that some types of cultural group, called 'communities of practice', do exist within organisations (Brown and Duguid, 1991; Wenger, 1998), and that boundaries between them do inhibit the diffusion of innovations (Ferlie et al., 2005). In many organisations workers tend to interact best with those who share their practices, values, roles and interests. In a hospital, for example, communities of managers, doctors and nurses can be identified (Dopson and Fitzgerald, 2005; Ferlie et al., 2005). Knowledge in the form of good ideas for improvements could spread well enough within a community of practice, but tend to 'stick' at the professional boundaries. This is the 'knowing–doing gap' (Pfeffer and Sutton, 1999, 2000): a member of one community having a great idea for improvement, can make members of another community aware of it only to find they then fail to turn that knowledge into action. Conversely, employees who share practices with members of rival organisations, forming a 'network of practice' with them, may find it relatively easy to share ideas with these like-minded people. Thus knowledge sticks *within* organisations and leaks *between* them (Brown and Duguid, 2001).

In order to represent this scenario, Watts (2009) drew upon Collins's sociological theory of interaction rituals (Collins, 2004). When participants in a social interaction become aware that they are focusing attention on the same ideas, objects or practices as each other, this generates a charge of 'emotional energy'. This emotional attachment then influences people in their choices between future interaction opportunities. However, emotional charges decay over time unless renewed in further interactions. Hence, if novelty is to be introduced to a particular, mutually reinforcing group, they must first be deprived of opportunities to interact so often, or at least deprived of the props required for focusing on the old objects and practices. This suggests that temporarily isolating a few members of a group might be a useful first step towards convincing first them of the merits of the innovation, and then allowing them back to spread it to the rest. It also suggests that frequency of interaction is an important part of cultural dynamics, and technologies that enable people to interact more frequently, and with greater targeting of like-minded individuals, will lead to the emergence of different cultural groups. Such a theory might explain how the Internet Age has coincided with the emergence of terrorist and religious extremist groups who can spread around the globe without enjoying a large membership.

To sum up this section, organisational and bibliometric data reveal academic researchers to be forming clusters in their work, but simulation models can demonstrate a variety of explanations for such patterns. A preference for similarity forms the basis of several such models, but invites the introduction of further principles to moderate its effects on minority influence, group formation and divergence, and diffusion across cultural divides. Importing more research from social psychology and sociology may help enrich these models, including work on group dynamics and interaction rituals.

However, all of these suggest that much can be explained without introducing consideration of a world outside the social interactions and their participants. In particular, it is not clear what the simulated academic papers are about, and how this content might play a role in the structure of academic fields. It is time to turn to models of academic production in which contents or subject matter really matter.

SCIENCE AS SEARCH AND THE DIVISION OF RESEARCH LABOUR

Chapter 4 examined models of learning in organisations, in which learning and problem solving were conceived of as a form of search activity. It was noted that in performing searches the agents in these organisations could employ some short-cut rules of thumb, or heuristics, instead of evaluating every possible solution or combination of ideas. Science can also be likened to heuristic searching, with Sandstrom (1999, 2001) likening scientists to ants foraging for food in that they leave trails for other scientists to follow, who in turn enhance those trails (a cumulative advantage process applied to citations). The idea that scientists resemble creatures of very limited intelligence and no individual awareness of the wider world may seem a little strange. But in other contexts there is no denying the power of heuristics for reaching reasonably satisfactory solutions given vast search spaces. So if, in the spirit of Ockham's razor, explanations for scientists' progress should not be made more complicated than necessary, then the potential for modelling scientists as components of heuristic search mechanisms should be investigated.

Computer scientists and operational researchers have already identified the power of a heuristic search algorithm inspired by observations of ants. Ant-colony optimisation (Corne, Dorigo and Glover, 1999; Dorigo and Blum, 2005; Dorigo, Maniezzo and Colorni, 1996) involves generating candidate solutions from a set of components, evaluating them, and then using the results of the evaluations to weight the sampling of components for future solutions, so that components that led to particularly good solutions are more likely to be repeated. The adaptation of these weights, analogous to the laying down and decay of ants' pheromone trails, is a form of stigmergic learning: the current explorers alter the environment in which future explorers make their decisions. Similarly scientists are able to use past academic literature and conference presentations, as well as personal communications, such as from teachers and mentors, to guide them in their search for new knowledge. These sources represent part of scientists' adaptable environment.

Experience with other heuristic search algorithms might suggest further analogies. In Chapter 4 we discussed the idea, illustrated by March (1991) in the context of models of organisational learning, that there is a danger that one decides on a particular solution too soon, without exploring widely enough the other possibilities. This is the problem of premature convergence on a local peak or optimum, better than neighbouring solutions but not the best solution in the entire search space. Avoiding the problem seems to require striking a balance between *exploring* the search space further and *exploiting* what one has already discovered. In simulated annealing (Kirkpatrick et al., 1983) a search mechanism akin to hill climbing is modified to allow some inferior solutions to be accepted. The chances of this occurring decay with time and also depend on how inferior the new solution is. Likewise, science might escape a local peak if editors and reviewers underwent periods of greater open-mindedness as to whether the position advocated by a submitted paper was going to lead on to improved work. In tabu search (Glover, 1989, 1990) previously tested solutions are added to a tabu list, excluding them from further evaluation and forcing later searches to explore more widely. Academics' insistence that papers make original contributions to knowledge serves a similar function. Particle Swarm Optimisation (Clerc, 2006; Clerc and Kennedy, 2002) involves a division of labour between search agents. Manager or memory agents record the current best solution found so far. Their worker agents explore near variations on the recorded solution. If a better one is found, the manager now remembers this instead. To avoid premature convergence, the agents can be divided into 'teams' or 'tribes' (Clerc, 2006; Jin and Branke, 2005; Poli et al., 2006), each with its own manager and competing for workers who may switch teams in response to perceived successes. Similarly scientists can also operate in teams, within which professors play a different role to their assistants and PhD students. Team leaders compete for human resources – the brightest, most hard-working of students – and different teams may choose to focus on different research areas. Even between professors there may be some role distinction, with some recognised as authorities, able to introduce or name some novelty as 'knowledge', while others struggle to find acceptance for their innovations and concentrate instead on interpreting and teaching what the authorities have said. We showed in the previous section that a process based on homophily, or the preference for similarity, was one way to achieve this division of roles.

Contained within several of these analogies is the idea that the search activities of different knowledge workers affect each other. What are the implications of this? If multiple approaches to searching exist among a population of scientists, can they benefit from each other, even when employing different approaches? Is there an ideal approach to searching – an optimal heuristic – that all might follow, or do they benefit from there being multiple search methods within the same scientific field? To promote thinking along these lines, some philosophers have taken up computer modelling in what might be called 'computer-aided social epistemology' (Hegselmann and Krause, 2006). Consider again the opinion dynamics model of Hegselmann and Krause (2002), in which agents adjusted their beliefs under the influence of other agents whose beliefs had sufficient proximity to their own (*HK_OpDynBC.nlogo*). Suppose some agents, perhaps a minority, are also influenced to some degree by a direct perception of how things really are – the truth. A simple way to represent this is to make these truth-seeking agents set their next opinion to a weighted sum of the true value and the mean opinion among the agents in their confidence bound. In addition to the confidence bound size, Hegselmann and Krause (2006) demonstrate that the behaviour of this model varies according to the balance between truth-awareness and social influence, and also according to the number of truth-seeking agents in the population. Figure 5.9 shows an example population using our NetLogo replication of their 2006 model. In this example run, with confidence bound set to one tenth of the opinion scale, 10 per cent of the population are truth seekers and are influenced 50 per cent by the truth (set to halfway up the plot) and 50 per cent by the agents in their confidence bound. Over time, the truth seekers migrate towards the true opinion, and, due to their social influence over others, some of the truth-unaware agents are drawn by them to follow. The more truth seekers there are spread across the opinion range, the more likely it is that an unaware agent will be influenced by one. But if the truth seekers adapt too fast, their influence will be short-lived. Truth-unaware agents will be left behind and converge on consensus positions remote from the truth. As the models of organisational learning demonstrated in Chapter 4, agents can gain knowledge collectively that individually they would have missed, but their collective performance depends on getting the balance right between different principles or influences.

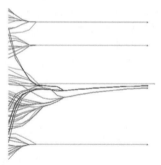

Figure 5.9 *In a mixed population of truth seekers and truth-unaware
            agents, social influence can lead to the latter agents following
            the former towards the true opinion. If truth seekers move too
            fast, however, truth-unaware agents can be left behind.*

Moving on from the one-dimensional world of Hegselmann and Krause's
model, Weisberg and Muldoon (2009) have proposed a simple illustrative
model of search on what they call an 'epistemic landscape'. The idea that a
landscape is a good metaphor for the search space of all possible solutions
or combinations of ideas and practices was introduced in Chapter 4. In
Weisberg and Muldoon's model a research topic is represented by a single
epistemic landscape. The dimensions of this landscape represent different
decisions about the approach to the topic. Decisions to be made might
include the research question being pursued, methods of data collection,
methods of data analysis and theories by which to interpret the analysis.
For each decision, there may be multiple options. Each approach yields a
value, which we shall call its 'fitness', but which Weisberg and Muldoon
refer to as the 'epistemic significance' of the approach. What determines
the value of a particular approach – whether it stems from political and
social influences, such as dominant ideologies, or from technological and
material influences – we shall, like Weisberg and Muldoon, leave unan-
swered. The fitness/significance values of each approach are set at the start
of the simulation and remain fixed whatever happens during the run.

Our NetLogo replication of Weisberg and Muldoon's model is available
on the website (*WMDivCogLabour.nlogo*). For simplicity of programming, and to communicate their points clearly, they use a visual representation of landscape search, with the two onscreen, spatial dimensions
representing two decisions to be made about each approach. Scientist
agents taking different approaches then appear at different coordinates.
Weisberg and Muldoon's choice of landscape is shown in Figure 5.10. Two
two-dimensional Gaussian functions are summed to produce a relatively

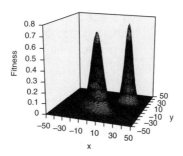

Figure 5.10   A simple fitness landscape with two 'hills'. The fitness function is here based on a combination of two Gaussian functions in two dimensions.

simple landscape with two 'peaks', one slightly taller and the other slightly fatter. Patches some distance away from either peak become indistinguishable from their neighbours, creating wide areas of positions one would be indifferent between. At initialisation, all agents are given random positions among these patches. With no clues in their immediate neighbourhoods about the way to better patches, the agents will be forced to move some distance about the landscape to find any improved positions.

The fitness value of a particular approach can only be known if an agent uses it, i.e. visits the corresponding patch. However, agents can detect whether any of the patches in their neighbourhood, the eight patches around them, have been visited by any agents. This information about visits can be used to decide which patches to try next. Three types of agent are compared, called 'Controls', 'Followers' and 'Mavericks', each with their own search behaviour. The ideas behind these respective types are as follows. *Controls* engage in a form of trial-and-error experimentation, a version of the heuristic known as *random-walk hill climbing* introduced in Chapter 4. Both *Followers* and *Mavericks* make use of the fact that previous searches have altered the environment in a way they can detect and learn from. The decision rules governing their behaviour are roughly as follows.

Controls:
Is my current fitness better than my previous fitness?
        Yes: Take a step forwards
        No: Is current fitness equal to previous fitness?
                Yes: If lucky (2 per cent chance) choose a random heading and take a step forwards, else do nothing.

No: Take a step backwards and choose a random heading.
Update record of previous fitness.

Followers:
Is there a visited patch in my neighbourhood?
No: Take one step towards a randomly chosen neighbouring patch.
Yes: Are the best visited patches in my neighbourhood better than my current fitness?
Yes: Take a step towards one of the best visited patches in my neighbourhood.
No: Is there an unvisited patch in my neighbourhood?
Yes: Take a step towards one of the unvisited patches in my neighbourhood.
No: Do nothing.
Update record of previous fitness.

Mavericks:
Is my current fitness greater than or equal to my previous fitness?
Yes: Is there an unvisited patch in my neighbourhood?
Yes: Is the patch immediately ahead of me unvisited?
Yes: Take a step forward.
No: Take a step towards one of the unvisited patches in my neighbourhood.
No: Is there a patch in my neighbourhood with fitness greater than or equal to my current fitness?
Yes: Take a step towards one of the patches in my neighbourhood with fitness greater than or equal to my current fitness.
No: Do nothing.
No: Take a step back and choose a random heading.
Update record of previous fitness.

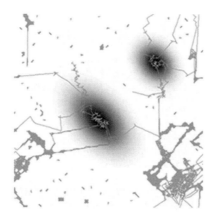

Figure 5.11   When a Maverick, taking mostly straight paths, finds a peak, Followers can reach the peak by travelling along the Maverick's path.

A typical simulation run of 500 time steps, with 100 Followers and 10 Mavericks, is shown in Figure 5.11. As can be seen here, by being able to sense which patches have been visited, Mavericks and Followers are able to interact. In particular, the Mavericks are best at exploring the landscape, and thereby the agents who reach peaks first are most likely to be Mavericks. But Followers can use the paths of visited patches traced by Mavericks to reach the peaks also. Otherwise, Followers tend to explore thoroughly a relatively small area of patches. On a landscape with multiple peaks located closely together this behaviour might be useful in locating the best, but on this landscape it has resulted in too many Follower agents exploring what are uninteresting regions.

How can one get Followers to more interesting parts of the landscape? The answer is to introduce some more Mavericks. Figure 5.12 shows results from populations of 400 agents with varying proportions of Mavericks. The higher the proportion of Mavericks, the easier it is for Followers to find paths to better fitness, but the worse Mavericks do. Mavericks try to avoid already-visited patches, including those visited by other Mavericks, and so can obstruct each other's passage to peaks.

Weisberg and Muldoon's model illustrates the advantage of there being people who do their own thing, seeking original approaches, but also the advantage a mixture of search behaviours in a population of scientists. 'Scientists' micro-motives can look epistemically impure or short-sighted, yet these motives can actually help the community as a whole make rapid progress towards finding out the truth ... [W]hat is epistemically good

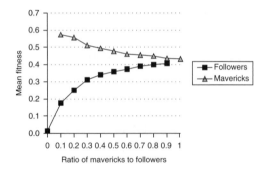

*Figure 5.12   Followers benefit from the addition of Mavericks, but
                Mavericks can obstruct each other's passage to fitness peaks.*

for individuals may differ from what is epistemically good for the community.' (Weisberg and Muldoon, 2009).

How best to explore the approaches to a topic depends on how interdependent rival scientists are. It also depends on the topic, represented
in this model by the shape of the epistemic landscape. As with the first
hill-climbing example in Chapter 4, this example landscape was simplistic, with just two decisions to make, and resulted in a relatively
easy-to-explore two-peak evaluation of the approaches. Real scientific
topics involve multiple decisions and may require much more difficult
search tasks. Sensitivity to the landscape's ruggedness, or search task
difficulty, could be tested using different landscapes such as Kauffman's
NK fitness landscape (Kauffman, 1993, 1995, 2000) described in Chapter
4. Scharnhorst (1998, 2001) suggests real knowledge landscapes might
be estimated from analysis of bibliometric data. Topics and sub-fields
for which the rate of publication was high might imply relative ease of
discovery – although they might alternatively indicate the application of a
relatively large search effort. Without data on the resourcing required for
the publications, it is difficult to see how an inference can be made from
the number of successes. Things become more complicated still if we allow
that scientists may be influenced by more than how well their beliefs and
actions 'fit' some external, material reality. The models proposed earlier
for explaining clustering among scientists and papers assumed scientists
were influenced by considerations of similarity or familiarity. It is also
an intended requirement for academic papers to be original, or not too
similar to previous work, though unintentional repetition is not always
identified by authors and reviewers. The challenge for a would-be academic author is to find methods that work reliably well when tested and
ideas that fit well together and resonate for other readers, but which are

sufficiently novel for reviewers and editors to deem them original. If in addition they cause controversy among some readers, even better, for then other authors will be energised to write papers in response, increasing the citation counts for the first paper and attracting further readers. All this makes the task of successful publication seem complex indeed, though it probably varies from field to field. Whitley (2000) suggests academic fields differ in their task complexity and in the degree of interdependence among authors, with the result that different organisational structures emerge among the academics working within them.

To conclude this section, the activities of academic scientists contain multiple analogies to heuristic search mechanisms, and incorporating these in simulations might allow us to explore the impact of promoting different approaches to scientific research. How good the analogy between model and real-world science is may depend on how we represent the task or landscape the scientists have to explore. The design of this is no trivial matter, but the ability to draw lessons for science policy may depend on it.

TOWARDS A COMBINED MODEL OF SCIENCE

So far we have described models of cumulative advantage processes, capable of generating scale-free frequency distributions for papers per author and citations per paper, models of opinion dynamics and cultural group formation, generating clustering among papers into subfields, and models of science as heuristic search on a landscape, generating 'fitter' combinations of ideas. Each of these processes picked up on a different aspect of academic science. The question now is, can the generative mechanisms be combined into one, more general, model of science?

Gilbert (1997) was the first science model to attempt this. As Meyer (2011) notes, this model reproduces a number of stylised facts about science. It generates a scale-free distribution of papers per author, in line with Lotka's Law (Lotka, 1926; Simon, 1955b). It exhibits the 'Matthew effect' (Merton, 1968) in that well-cited authors receive disproportionately more of future citations. It represents exponential growth in the numbers of papers and authors (Price, 1963). Finally, scientists cluster according to the contents of their work, resembling the 'invisible colleges' of specialities, which Price observed forming for approximately every 100 scientists.

Meyer (2011) suggests two more stylised facts for a science model to address. First, the timing of citations received by papers indicates a half-life effect (Burton and Kebler, 1960), with the chances of citation decaying over time. For some natural sciences, for instance, the half-life is about 5 years. Second, there is Bradford's Law, another scale-free distribution, governing the scattering of papers on a particular topic across journals (Bradford, 1985; Cole, 1962). This holds that for any particular field searching for papers on some topic within that field will yield exponentially diminishing returns. To date no explanation for such a pattern among journals has been identified. In addition, attempts to confirm Bradford's Law have been found to be problematic (Hjorland and Nicolaisen, 2005; Nicolaisen and Hjorland, 2007). But if true, then there are a number of practical implications. For librarians, space can be saved by identifying and stocking only 'core journals' in each field, since it seems little will be lost if more peripheral journals are omitted. For researchers, focusing on the most central journals would appear to save time. But for authors wishing to be read and cited, there is considerable incentive to get into the core journals, even though the increased competition and limited space for papers makes it less likely they will succeed.

In addition to these stylised facts, network analyses of co-authorship, co-citation and other bibliometric-based relations suggest more patterns to be explained. The TARL model ('Topics, Aging, Recursive Linking') of Boerner et al. (2004) simulates the evolution of paper and author networks, as well as fitting some of the other stylised facts. In particular, papers are given 'topics', in which they match or differ, and ages, which determine their chances of being read and cited, similar to the 'recency' concept in the cumulative advantage mechanism described earlier.

None of these models, however, incorporate the problem-solving or search aspect of science. We have attempted this with our *CitationAgents* model, which is capable of generating several of the stylised facts, and capable of performing heuristic search on a landscape, although combining the two types of capability seems problematic (Watts and Gilbert, 2011; see the website for Excel and NetLogo versions of the model: *CitationAgents_NK.xls* and *CitationAgents_NK.nlogo*).

Some of the mechanisms are familiar from *PapersPerAuthor.nlogo*, described above. There are two types of agent, authors and papers, within some academic field. The number of opportunities to write papers can be set to grow exponentially over time, and fit with bibliometric data from a real journal, which we illustrate with a journal from the field of innovation studies, *Research Policy*. A given probability distribution controls the number of authors per paper, for which we use a Weibull distribution to fit the data. Each author may be sampled from those of past papers, or may be a new entrant to the field. A rate of new entries is chosen such that the number of authors grows exponentially over time to fit the pattern observed in the data. Existing authors are sampled in such a way that cumulative advantage operates, and the number of papers per author tends towards a scale-free distribution. Papers also have references to previous papers. The number of these is sampled from another Weibull distribution derived from data. Watts and Gilbert (2011) test several mechanisms for sampling papers to be cited, with some mechanisms involving cumulative advantage and able to generate the scale-free distribution for citations per paper. Age-based Weibull-distributed weighting functions are also applied to the past papers used for generating authors and references. Thus the half-life effect can be approximated, and other time-based data patterns also fit well.

Papers have contents, and authors have beliefs, both of which have a 'fitness value'. Watts and Gilbert (2011) represented contents and beliefs as bit strings, in common with the models of organisational learning in Chapter 4. The *NK* fitness landscape model (Kauffman, 1993, 1995; Lazer and Friedman, 2007) is used to define a value for each paper's contents and each author's beliefs. With $N = 20$ bits and $K = 5$ interdependencies, the field is a moderately difficult search problem. Peer reviewers, sampled from the authors of past papers, decide whether a new paper is to be accepted or rejected for publication in a journal. Given a distinction between publications and unpublished papers, the sampling mechanisms for authors, references and reviewers can optionally be restricted to publications only. Peer reviewers vote against publishing a paper if its contents' fitness value is below their own beliefs' fitness, or below the value of its referenced papers' contents. Publications also had to seem to be original; reviewers also vote against anything identical to their beliefs or referenced papers' contents. Papers were only published if the number of peer reviewers voting for publication met some threshold value, in this case three.

Given contents and authors, papers could also be tested for similarity. When a first author sought co-authors for a new paper, a preference for similarity was defined in terms of the number of bits (beliefs) they had in common. Reference papers could be rejected for lack of similarity to the first author's beliefs (as if he or she could not understand them), and likewise peer reviewers could reject the final paper if they found it lacked similarity to their beliefs. Such principles should be sufficient to generate clustering among authors and papers, but Watts and Gilbert (2011) do not investigate, instead leaving all requirement thresholds for similarity set to zero bits. Inspired by Gilbert (1997) and Weisberg and Muldoon (2009), however, we have converted the model to represent beliefs and contents as positions in a two-dimensional space. (*CitationAgents_2DLandscape. nlogo* on the website.) Originality and similarity can then be defined in terms of the Euclidian distance between two points. An alternative definition of fitness is required, for which we can borrow the double Gaussian landscape from Weisberg and Muldoon (2009).

Hence we can show visually publications progressing towards fitness peaks. We can also show evidence of clustering, in this case due to review-ers requiring similarity. Getting the model to show both forms of behav-iour, however, proves to be tricky, as the two preferences, for improved fitness and for similarity, can conflict. In addition, once papers can be rejected for publication, sampling methods for choosing authors, refer-ences and reviewers restricted to publications mean that it becomes much harder to generate the growth curves and frequency distributions desired for a science model, as we reported for the bit-string version (Watts and Gilbert, 2011). Compared to generating these patterns without the paper-publication distinction, the task has become much harder. The bibliomet-ric data from *Research Policy* was being used to identify distributions for *paper* generation, but it actually represents *publications*. The problem is, we lack data on papers that are written but do not get published and hence do not enter our empirical dataset. Information may be available from some journals concerning what proportion of submitted papers are rejected (for example, 10 per cent), but this does not tell us the fate of rejected papers. They may be, for example, resubmitted to other journals, broken up and their material recycled in other publications including book chapters, or just left online as working papers, where they may still influence the field if someone reads them. Exploring how sensitive our models' behaviour is to changes in the generating mechanisms might give some clues as to which writing and reviewing processes would leave us with realistic patterns. But the parameter space is vast, so even automating the search for good data fits might not be much help. Ethnographic studies of how scientists create papers and how peer reviewers evaluate them may be required to give more insight into the mechanisms we should be representing, and thereby reduce the difficulty of a realistic science model.

The question we addressed in Watts and Gilbert (2011) was 'does cumu-lative advantage affect collective learning in science'? Organisational learn-ing models demonstrate the danger of premature convergence on inferior combinations of ideas and practices, and the need to balance *exploration* against *exploitation*. Cumulative advantage, focusing authorship oppor-tunities on just a few authors, and citations on just a few past papers, would restrict how widely a field is explored. Preferences for similarity would have the same effect. We found negligible differences between sam-pling with and without cumulative advantage. In contrast, sampling from peer-reviewed publications rather than all papers made a noticeable dif-ference to the collective performance of scientists as searchers. However, as we have already noted, the model ceases to appear well-calibrated and plausible in its output, once the publication-paper distinction has been introduced. Recent trends concerning how academics are resourced and incentivised suggest that the question may be worth pursuing further, however. Various government-backed developments seem aimed at preventing inferior research being included among that funded – what statisticians call type I errors (Gillies, 2007, 2010). But in so doing, they risk errors of exclusion (type II errors), by excluding projects from funding that might have gone on to make valuable contributions if they had been allowed to continue a little longer. To echo Lord Leverhulme's famous quote concerning his advertising budget, x per cent of research funding and human resourcing is wasted, but the problem is we do not know which x per cent it is. In trying to judge this, we will make errors of inclusion and exclusion. Many academics are pessimistic about which way the error rates are going. If academic employers insist on their staff targeting their submissions on only highly cited journals, or journals rated as '4-star', there are likely to be a number of undesirable outcomes. The targeted journals could accept more papers for publication. But limitations on how many of these one can read and interpret mean that informal selection processes must still occur (Thorngate, Liu and Chowdhury, 2011). Alternatively, the journals could attempt to raise their standards, and reject more submissions. However, making reliable distinctions at this level will be difficult, with the risk that many excellent papers are rejected and thereby condemned to be submitted to lesser rated and read journals, assuming their authors' grants and jobs are not cancelled before then. In trying to identify more papers for rejection, reviewers may turn to relying more on familiarity to judge acceptability than on trying to identify quality or forecast whether the paper is going to represent a building block to major discoveries in the future. Agent-based models of science as search could make more explicit the idea of a trade-off between inclusion and exclusion errors, exploration and exploitation. When thinking about the productivity of innovators there is a tendency to believe that cutting resources means higher productivity. If at the same time there is a fall in production (number of novel ideas generated, number of problems solved etc.), productivity, or the ratio of production to resources, may fall instead of rising.

CONCLUSIONS

As was said at the beginning of this chapter, studying science seems to be an excellent way to study innovation. Scientists generate novelties, through novel combinations of existing ideas, techniques and places to explore. Through teaching, mentoring, presentations and publication there is diffusion of these innovations, and new developments are built upon previous ones. The study of science means the study of scientists, and it is clear that innovation is a social phenomenon.

Scientists' publications data represents a valuable empirical trace of innovation activities, and a role for social simulation models was found in testing whether simple mechanisms could explain how certain patterns come to be present in the bibliometric data. These stylised facts include exponential growth in numbers of papers and scientists, and scale-free distributions for the numbers of papers per author and citations per paper. We presented a program for generating the latter using a process of cumulative advantage, and noted that this implies that a prolific or well-cited author is not necessarily a more talented one. We also learnt from an opinion dynamics model and Axelrod's model of cultural influence, how another simple mechanism, based on preference for those who are similar to us, could explain the emergence of clusters among scientists and their papers. So something endogenous to scientists' activities might explain what others have said stemmed from factors external to them, such as sources of resources or even just how reality is. However, if the acceptability of an author's argument depends on how similar or familiar it seems to audiences, editors, reviewers and readers, then it is conceivable that especially novel ideas might be overlooked, irrespective of whether they described more effective or efficient ways to observe, experiment or produce. The careers of overly adventurous authors might suffer if science was too conservative.

Finally we noted several aspects of science that resembled heuristic search methods. This promised a partial explanation for how science could make progress in seeking novel solutions to problems, or new combinations of ideas that seem to fit the world better. The explanation is only partial, however, since we need to know something about the type of search problems that scientists face. If search in science is analogous to exploring a landscape, the modeller needs to know more about how to represent the landscape's structure. But given a landscape definition, it was easy to demonstrate the impact of different search practices and organisations of scientist searchers, including how the presence of 'mavericks' might benefit those whose advances in science were more tentative, gradual and thorough.

There seemed some hope for simulation models of science to guide our thoughts when thinking about science policy, such as the allocation of resources, the location and organisation of scientists and their institutions, the provision of opportunities for communication, and the structuring of journals and their peer review system. If science were reorganised in some way, could new ideas be had faster or at lower cost? Are some ideas being missed because scientists do not explore widely enough? As with the models of organisational learning, we might expect a trade-off between exploration for new ideas, and exploitation of those we have already.

But we face a hard task in producing a science simulation that addresses such questions while still appearing to represent what really happens among scientists. Each of the generative mechanisms surveyed in this chapter has the potential to interfere with the action of another.

The bibliometric data leaves many things out: the amount of resources contributed to the production of each paper, what happens to unsubmitted and rejected papers, and why people decided to co-author, reference a particular paper, or accept one for publication. Ethnographic studies of how science works can help here, and simulation modellers of science need to incorporate their findings as well as learning from analyses of bibliometric data. Agent-based simulations can then help identify formal descriptions of the mechanisms and spell out the macro-level consequences of these micro-level behaviours.

This is one development for future science models. In addition, there needs to be more work on combining processes and testing sensitivity between them. Network analyses of co-authorship, co-citation networks and science maps based on keywords and contents may yield more clues. Finally, we might note that although science simulations are growing in number – see for examples, a recent *JASSS* (*Journal of Artificial Societies and Social Simulation*) special issue (Edmonds et al., 2011) and a book (Scharnhorst, Boerner and Besselaar, 2012) – to date none of them consider the role played by teaching. Teaching not only determines what concepts the next generation of scientists have in mind to work with, but also influences the status of scientists outside academia. Indeed, by continually breaking down the barriers between the knowledge-haves and knowledge-have-nots, teaching forces academic scientists to keep searching for novelties, so as to restore – temporarily – their apparent specialness, a process Fuller (2004) has dubbed the 'creative destruction of social capital'. Future science models could address how the activities of scientists interrelate through innovation with the wider society.

6. Adopting and adapting: innovation diffusion in complex contexts

INTRODUCTION[1]

Models are 'tools for thinking' (Pidd, 1996), and different tools may inspire different thoughts, and lead to different policies. Experts in two different modelling approaches may arrive at different questions and identify different issues when confronted with the same scenario (Morecroft and Robinson, 2006). How the processes of innovation have been modelled then ought to be of concern to innovation studies. Chapter 2 surveyed the major types of model of innovation diffusion, including the epidemic and probit models. In this chapter, a new simulation model of the diffusion of innovations is presented, one in which some of the assumptions of other diffusion models are challenged. In particular, the distinction between the innovation *generation* and *diffusion stages* is weakened in favour of a view of innovation adoption as involving adaptation of the innovation. This adaptation is necessitated by the local variation in individual adopters' contexts; no one innovation satisfies all potential adopters. Drawing upon ethnographic studies of the social construction of technology (SCOT) and actor–network theory (ANT), it is proposed that diffusion of innovations should be modelled as a form of constraint satisfaction. But in allowing adaptation of an innovation, one reduces the importance of its originator: later, adapted versions may prove more popular than the original. Hence, there are likely to be practical implications of the thinking represented in the new simulation model.

To challenge some of the assumptions behind the so-called *linear model* of innovation is certainly not new (Balconi et al., 2010). However, the continued popularity of the traditional models suggests that alternatives have lacked persuasiveness. Part of the explanation for the popularity, we suggest, is the visual appeal of the traditional models. The new simulation model is intended therefore to communicate its points in visual forms, something agent-based modelling is particularly suited for. Another part of the explanation is the link between traditional models and the use of quantitative data sources. A more complex view of innovation is harder to derive from quantitative data, but qualitative research has indicated the need for such a view. A number of methodological points, therefore, are raised by the use of the new simulation model.

In what follows, it is explained in more detail why a new approach to modelling the diffusion of innovation is called for, and how the new approach is to be constructed. The new simulation model is described, along with some example results. A final discussion covers the methodological issues and the use of the model.

THE NEED FOR A NEW MODEL OF INNOVATION DIFFUSION

Competing Diffusion Models with Rival Lessons

As stated already, different models inspire different thoughts, which are then expressed in different policy actions. Following Geroski (2000), Chapter 2 discussed the main types of diffusion model and the contrasting aspects they each focus attention on, which we briefly recap here.

In *epidemic models* adoption occurs as a result of some kind of transmission process between existing adopters and new adopters, such as imitation or word-of-mouth advertising. Thus the focus is on social interactions, and managerial interventions to increase diffusion might include increasing the rate of interaction, and restructuring the network of interaction possibilities, such as directing people to talk to key opinion formers, and encouraging social contacts between people in separate buildings, groups or organisations.

By contrast, in *probit models* potential adopters may have access to the same information about the innovation at every point in time, but differ in how they react to it. Adoption decisions are made and remade in the light of new information, but decision-making entities are heterogeneous in their properties, and consequently the same new information, such as a fall in the price of adoption or an increase in the known capabilities of the innovative product, may persuade some to adopt and others to wait. The probit model then focuses attention on the properties of would-be adopters, and the dynamic properties of their environment, especially the economic properties. Those hoping to use this model seek suitable datasets, such as the sizes of potential adopter firms, and suitable theories of the decision environment, such as the relation between firm size and learning (Davies, 1979).

Further diffusion models include *stock* and *evolution models*, which focus attention on the effects of past and current adoption levels. Potential interdependencies between innovative products, their competing products, and other products and services may affect the costs and benefits in adopting. For example, a pool of adopters can represent a market opportunity for support services, and, as suppliers of those services increase, the supported product becomes more appealing, thus leading to further adoption and greater demand for support services. Positive feedback loops like these can turn a small advantage over rival products in adoption levels into later dominance of the market, and create legitimation as the standard other products have to be compatible with (Arthur, 1989). The system is *path dependent* (Arthur, 1994; David, 1985, 2001): each adoption decision changes the conditions in which later decisions are made, but early adoption decisions will wield more influence than later ones over future adoption and the final distribution between rival products. So managerial thinking should focus on interdependencies and network externalities, but also on timing and the advantages of being first to market. Indeed, in the model of information cascades (Bikhchandani et al., 1992, 1998), explored in Chapter 3, decision makers use past adoption decisions by others as evidence for or against their own adoption. In such a scenario, the earliest decision makers wield great influence over the later ones, who follow as a herd whichever adoption action reaches first sufficient popularity to appear to be the consensus. This type of diffusion model focuses attention on the potential for fads and fashions and bandwagon effects, not only important to the product launchers who may benefit or suffer from whether their product becomes fashionable, but also important to those who may mistake overwhelming popularity for an indication of superior quality rather than just a lucky break in early adoption.

There is scope for combining elements of these contrasting models of diffusion. The information about the economic environment that decision makers employ in the probit model could reach them via word-of-mouth channels, for example. But this would complicate data collection, model fitting and interpretation of the model, and so, to date, hybrid approaches have not appealed. The epidemic model is a mainstay in the innovation literature in sociology, marketing and business school textbooks, Rogers (2003) being the classic textbook, while the probit model appeals more to economists, for example, Stoneman (2002). As a result, these different literatures are set to advocate different data gathering activities and different managerial interventions. If types of innovation vary in their susceptibility to these rival approaches, one might also expect sociologists and economists to differ in what kind of innovations they typically study.

THE SIMPLIFYING ASSUMPTIONS OF TRADITIONAL MODELS

A cornerstone of traditional diffusion models is the concept of the diffusion curve: that the total number of adopters will form an S-shaped or logistic curve over time, while the rate of adoption tends towards a bell-shaped curve (Rogers, 2003, p. 298). The appeal of this concept is easy to identify: it communicates the phenomena of diffusion in an elegant visual form, the mathematics behind it is relatively easy to understand (as differential equations go) and it seems to fit the adoption data from the classic study of the diffusion of hybrid seed corn among mid-west US farmers (Griliches, 1957; Ryan and Gross, 1943). If one uses the most basic of epidemic models, the S–I disease model (Kermack and McKendrick, 1927), quantitatively the fit is not as impressive as it might first seem. Compared to the model, there is a time lag before the empirical adoption rate takes off, and the empirical peak rate is higher than the model predicts. As we argued in Chapter 2, it is not difficult to improve the fit with enhanced models: for example, building in time lags to adoption, extrinsic sources of marketing influence, as found in the Bass (1969) model, and adding threshold effects, whereby decision makers only adopt after multiple neighbours have adopted (Centola and Macy, 2007). But now the problem is that there are a variety of rival models to explain the adoption data, resting on contrary assumptions, and the data fails to decide between them all.

The elegant, simple epidemic model achieves its logistic curve while assuming all decision-making agents are homogeneous in their susceptibility to the attractions of the innovation. Which agents adopt earlier than which others is the result of pure chance, and the order could be expected to change for another innovation's diffusion. If agents differed sufficiently in their susceptibility, the resulting bell curve for the adoption rate would be asymmetrical, or skewed. By contrast, studies of the personal characteristics of adopters found it useful to divide adopters into categories based on when they adopted (Rogers, 2003 Chapter 7). For example, 'Early Adopters' differed from 'Laggards' in social, economic and educational properties. This seems to support the probit model's assumption of agent heterogeneity, over the use of the simple epidemic model.

Another key assumption, behind the use of both epidemic and probit models, is the separation between, on the one hand, innovation generation and introduction to a population, and, on the other hand, the later diffusion of that innovation throughout the population. What diffuses is something fixed and immutable, the work involved in its creation having been completed by then. This is part of the so-called 'linear model' of innovation, whereby innovation is divided into various stages, each stage preceding and feeding into the next: basic research, applied research, development, production, marketing, diffusion. Criticism of this way of thinking about innovation is extensive. Indeed, the linear model has arguably become little more than a 'straw man' (Balconi et al., 2010). Its origins partly lie in an attempt to justify the funding of basic scientific research (Godin, 2006), and university research does seem to have fed into the biotech and pharmaceutical industries that have grown during recent decades. But it is clear that in other areas an innovation's success depends heavily on processes that occur after its introduction, in particular learning by using it (Rosenberg, 1982, Chapter 6).

Ethnographic studies from the field of science and technology studies (STS) bear this out. Seeking explanations for the relative failure of technologies to diffuse, they identify that products designed in one locale – French technology labs, for example – do not suit well the conditions found in destination locales, such as African villages (Akrich, 1992; Akrich, Callon and Latour, 2002). The would-be successful innovation 'must be transformed, modified according to the site where it is implemented. To adopt an innovation is to adapt it: such is the formula which provides the best account of diffusion.' (Akrich et al., 2002, p. 209). But this means that the originators of an innovation play a less important role than before. '[T]his adaptation generally results in a collective elaboration . . .' The instinct of designers of consumer goods is generally to enclose products in smooth plastic cases, to exclude the user's intervention in the internal workings, and to constrain what can be done with the product (Woolgar, 1991). By contrast, products which permit a high degree of modification, such as the personal computer, Internet services and more recently the *iPhone*, have seen an explosion of user-developed applications, many of them unanticipated by the manufacturers of the original device but each increasing its value to adopters.

CONSTRAINTS AS SOURCES OF COMPLEXITY

Why does adoption have to involve adaptation? The literature on so-called 'actor–network theory' and the social construction of technology employs ethnographic studies of how technologies are developed and used. These studies reveal that the traditional models of generation and diffusion of innovations miss out important sources of complexity. A regular theme is that of both design projects and adoption decisions involving the *satisfaction of constraints*. Three classic case studies may suffice to illustrate the principle.

There are multiple examples of constraint satisfaction problems described in Latour (1996), an investigation into the causes of the failure of the project to develop a new public transport system in Paris, called *Aramis*. The system is to have some of the elements of automobiles, taxis and metro trains, but there are constraints between these. For example, the relatively small size of platforms and passing loops means it is difficult for cars to alter speed, which means a car leaving a platform will cause delay to other cars, which in turn means cars must travel slower, and hence there will be less traffic capacity. To have longer platforms and passing loops, however, would mean there would be fewer stations and hence longer walking times for pedestrians between stations. This in turn would mean fewer customers. The *Aramis* project ran for many years, consuming plenty of funds, but between 1970 and 1987 only two minor changes were made to the specification list for the system (Latour, 1996, pp. 279–280). Delays and high costs can be blamed on the difficulty in satisfying so many constraints, but no one involved in the project seeks to reduce the list. There are political constraints on *Aramis* as well. Make it more like one of the existing modes of transport, and its novelty will be lost, and thereby the excitement it can generate. Simplify the technological challenge involved and the project will no longer provide a demonstration of French engineering prowess. Either way, the high cost compared to extending existing modes will not be justified without these properties. Thus in *Aramis* economic, political and technical constraints are interwoven.

There are plenty of political actors placing constraints on a project in Law and Callon's (1992) description of the failed UK project to develop the TSR.2, a new fighter-bomber aircraft in the 1960s. The list includes the air force (who will want a supersonic fighter and will want something superior to the other armed forces), the air ministry, the ministry of supply, the navy (who will want the aircraft to be dropped in favour of the air force buying the navy's own aircraft, Buccaneer), the treasury (who will just want to lower the cost), both Conservative and Labour Parties (who will oppose each other's policies) and the unions (who will want to protect jobs). The contractors submitting proposals have differing strengths: Vickers propose a short-take-off-and-landing aircraft; English Electric propose a supersonic plane capable of flying at low level. To preserve jobs in both firms, the government asks them to merge their designs. Rolls Royce offer the best quality engines, but they have already received other government projects, so Bristol Siddeley are asked to supply because they need the work. There are technical trade-offs also. Siting the engines in the fuselage would allow thin, clear wings, but will increase the risk of fire (and a fatal explosion did in fact occur during testing.) New problems emerge during the project's run. As the cost soars to £10 million per aircraft, the plane is now considered too unsuitable for risky missions. Meanwhile a change of government means the loss of political allies. Eventually, the TSR.2 is cancelled in favour of buying the navy's preferred option, Buccaneer, and buying American (the F111). The moral of the story, as told by Law and Callon, is that *local events*, such as the design decisions made by contractors or the political decisions made by ministers, have consequences on a larger scale, including the project's success or failure, and also the nation's defence capability.

Emergent effects are a theme of Bijker's (1992) account of the development of the fluorescent lamp. There are rival conceptions of fluorescent lighting: a tint-lighting fluorescent lamp versus a high-efficiency fluorescent daylight lamp. A compromise solution, a high intensity daylight fluorescent lamp, emerges from social interactions between two main groups: the Mazda companies who will manufacture it, and utility companies who want the technology to use plenty of their product, electricity. As such, the solution represents an artefact that was 'invented in its diffusion stage'.

The actors involved in these cases are heterogeneous in their interests, their interpretations of the innovation, and their degree of support for it. There are other technologies and practices in circulation which affect the innovation's value, and rival products may be adopted first, thereafter blocking its diffusion by using up available resources. Satisfaction of all these interests may not be feasible, and establishing the extent to which this is so may cost time and money. Development, like adoption in Akrich et al. (2002), involves adaptation, in this case compromises between actors. Diffusion models that assume homogeneous agents or independent products leave poorer one's attempts to understand the emergence or non-emergence of a successful innovation. A modelling approach that focuses attention on the existence of constraints might therefore prove more enlightening.

THE MODEL OF ADOPTING AND ADAPTING

Modelling the Satisfaction of Constraints

How best to model the satisfaction of constraints? A first response might be that some classic simulation models already do. In Schelling's model of segregation, there are constraints on agents of one 'colour' living next to agents of another 'colour', and agents consider moving to minimise the extent to which these constraints are broken (Schelling, 1971). Axelrod's model of cultural influence places similarity constraints on neighbouring agents' cultural features (Axelrod, 1997a Chapter 7; 1997b). Epidemic models of diffusion are themselves constraint satisfaction processes, as agents find dissatisfactory a state in which neighbours possess some item which they themselves lack, and consequently act so as to 'keep up with the Joneses'. These are all examples of social constraints, that is, constraints on the attributes of neighbouring people in some social or spatial network.

Constraint satisfaction problems more akin to the technical, material and resource-based ones mentioned in the three cases in the previous section are already tackled within operational research and computer science. They include scheduling work, choosing a location and allocating resources, but the simplest form is the *K-Sat* problem (Gomes and Selman, 2005; Kirkpatrick and Selman, 1994). In this, values are sought for V binary variables, which satisfy as many as possible of a list of C constraints. A constraint is a logical proposition involving K of the variables. One known result concerning this idealised problem is that both the difficulty of finding a solution that satisfies all C constraints, and the expected number of constraints that can be satisfied using a given amount of searching, depend on the ratio of constraints to variables (that is, on C/V) (Kirkpatrick and Selman, 1994). Indeed, as this ratio increases, the problems undergo a phase transition, from being feasible and 'easy-to-solve', to being impossible to solve in a reasonable number of steps, if not outright infeasible. Figure 6.1 illustrates this for a K-Sat problem, with constraints of length $K = 2$ and logical form $X \ OR \ Y$, and $V = 16$ variables. Twenty agents each seek a satisfactory solution using a simple heuristic search method, 'hill climbing', which we described in Chapter 4.

Kauffman (2000, Chapters 8, 9) has already employed K-Sat problems as a model for technological evolution, where technologies, by their presence or absence, constrain or enable the existence of each other. In Kauffman's model, a population either has or does not have a particular technology, represented by a single binary variable. The possibility of

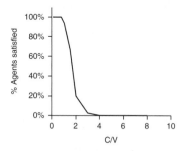

Figure 6.1   Phase transitions in the difficulty of K-Sat constraint satisfaction problems. As the number of constraints, C, divided by the number of variables, V, increases, it becomes impossible to satisfy everyone in the time given.

only some members of a population possessing or knowing the technology is not modelled. Diffusion is instantaneous and universal then. In the current chapter the idea is extended to a diffusion model with individual agents capable of differing in their current adoption decisions, given multiple technologies diffusing. The K-Sat model is used to represent the constraints on individuals' adoption decisions.

OUTLINE OF THE AGENT-BASED SIMULATION

How best to combine a model of diffusion with that of constraint satisfaction? We use agent-based simulation modelling to represent individual agents making adoption decisions and adapting technologies to their local needs. Examples already exist of epidemic models using this simulation approach (Bagni, Berchi and Cariello, 2002; Dunham, 2005; Rahmandad and Sterman, 2008). As we have seen in Chapters 2 and 3, such simulations allow the representation of heterogeneity in agent attributes, including positions in social networks. Agent-based simulations also typically include stochastic processes, that is, processes with random, variable outcomes. In systems susceptible to path dependence, chance local events such as the order and point of origin of rival innovations can cause major differences in longer term, aggregate outcomes. Alternative simulation models of diffusion, including system dynamics models (Sterman, 2000) and economists' equation-based analytical models (Stoneman, 2002), become cumbersome to design when representing even modest amounts of heterogeneity, and typically do not consider the breadth in outcomes from models with both stochastic processes and path dependence. Representing the effects of social network structure is also difficult in these approaches. Social networks may be important for modelling constraint satisfaction: the agent-based models of organisational learning in Chapter 4 have shown that social networks and the rate of problem-solving practices, such as trial and error and imitating others, can control the balance between exploration for new solutions and exploitation of existing ones (Lazer and Friedman, 2007). Agent-based simulations have to perform far more computations to generate population-level statistics such as the S-shaped adoption curve, but their ability visually to represent agent interactions in varied, spatial environments gives them an aesthetic appeal and a communicativeness of their own.

A description of *AdoptAndAdapt.nlogo*, the agent-based simulation of adopting and adapting technologies, now follows. A population of agents exists in some world. Each agent holds a set of items representing the technologies, ideas or practices he or she has adopted. For simplicity we shall refer to these always as technologies. An agent's set or portfolio of technologies is represented within the simulation as a string of binary variables, with '1' signifying that the corresponding technology is present in that portfolio, and '0' signifying that it is absent. Interdependencies between the technologies mean that agents experience constraints on which technology goes well with which. An agent's degree of satisfaction with its current portfolio is defined as the number of constraints currently satisfied by the current values of the portfolio's binary variables.

Agents seek to improve their individual satisfaction. To do this, each agent tries periodically one of the following actions: to invent or discover a new technology, to imitate another agent within a given radius (if any such agent exists with a technology to imitate) and to discard or forget a technology currently adopted. During invention, a randomly chosen technology that is not currently present in the agent's portfolio is added, providing such a technology exists. If the agent is not less satisfied with the new state of its portfolio, the invention is kept. Otherwise the invented technology is removed, and the old portfolio preserved. (This is in effect the heuristic search method of 'random-walk hill climbing'.) During imitation, the imitator copies one randomly chosen technology from the portfolio of the imitated agent, selected from all technologies possessed by the imitated but not by the imitator. Again, if the addition does not make the imitator's new portfolio less satisfactory, the addition is kept. During the discarding action, a currently held technology is chosen to be removed from the portfolio. If the resulting portfolio is more satisfactory than the old, then it is kept without the discarded technology. Otherwise the agent cancels the discard action. (It will be noted that when agents' satisfaction levels are the same with or without a particular technology, their preference is to opt for the portfolio with the larger number of technologies.)

The agents live in a spatial, physical environment. Each agent has its own home base, a location from which it never roams further than some given distance, the radius of its 'roaming circle'. At simulation initialisation, all agents start at their home bases, pointing in arbitrary directions. Thereafter, each agent moves forward one unit per time step. If an agent reaches the edge of the roaming circle, it turns round to face its home base, and then chooses a new direction, keeping within the circle. If it reaches the edge of the simulated world, it tries another arbitrarily chosen direction. When combined with imitation of agents within a given radius, this movement within limits is intended to simulate the transmission of innovations within a social network that has been generated primarily by who lives near to whom. The model agents do not move home, or make radical changes in location and hence social network. However, over time there may be some small variation in which neighbours are currently close enough to imitate or be imitated.

As we discussed in Chapter 3, how best to represent realistically a social network is an open question (Hamill and Gilbert, 2009), and better alternatives may yet be identified. In particular, the agents in this largely regular, two-dimensional spatial network lack relations with more distant agents, which might represent other social ties such as family who have moved away. They also lack other typical locations: where they have just home neighbourhoods, human beings frequent workplaces and shops where transmission of ideas is again possible. But for a simple demonstration of transmission within geographical space, the existing rules of movement suffice.[2] It may be noted that if the radius of the roaming circle is set to zero, then the network of imitation relations becomes equivalent to the 'social circles' network of Hamill and Gilbert (2009). If the imitation radius is set wide enough, then every agent can imitate any other, and the network becomes complete.

The landscape or environment of this world consists of different terrains. Each terrain places a different set of constraints on the technologies that can be satisfactorily held simultaneously by an agent on that terrain. Constraint lists are generated at the start and held constant during the simulation run. In a simple scenario to illustrate the capabilities of the model, the landscape has been divided into just four quarters, with a different terrain in each quarter. From the colour scheme employed one may think of the four terrains as grassland, forest, desert and water (Figure 6.2). As agents move around the landscape, it is possible that some may cross from one terrain to another. In this case the set of constraints placed on that agent's technologies will also change. Thus an agent may find that a technology portfolio that was satisfactory to some extent in one terrain now becomes more or less satisfactory in the new terrain.

Additional sources of constraints might include the type of agent adopting the technology, other agents in the neighbourhood of the adopter (what is called in the model *Social-Constraints*), and the material nature of the technology itself, independent of any contextual features

Figure 6.2   Agents in the four-terrain landscape.

(called in the model *Generic-constraints*). For this work we will focus on the terrain constraints, and leave exploring the others to the reader. We shall also omit consideration of why agents might want satisfactory technologies. Unlike the agents in *Sugarscape*, the classic agent-based model by Epstein and Axtell (1996), our agents experience neither birth nor death and satisfying basic needs such as food, drink and shelter is not modelled. This model just assumes technologies and constraints are important.

To model *adaptation* of a technology, the concept of *complex technologies* is now introduced. These consist of a given number of *basic technologies* as components, set by the parameter *Size-Of-Complex-Techs*. The processes of invention, imitation and discarding now occur for complex technologies instead of basic ones. However, with a given chance, an agent may during imitation 'tinker' with the complex technology, adding or removing a basic technology from its components. Thus, a process of adaptation can occur during an attempted adoption event.

Model Behaviour

The constraints on which basic technologies go well with which mean that the appearance of one technology can affect the value, and hence adoption, of another. Figure 6.3 shows the number of adopters for each of four technologies in an example simulation run. For this example we used 30 agents on a 33-by-33 bounded grid world, the 'Four Quarters' landscape type, and a set of 16 terrain constraints for each of the four terrains, with each constraint being of length $K = 2$. The chances of invention, imitation and discarding are respectively 0.00001, 0.5 and 0.1. The agents

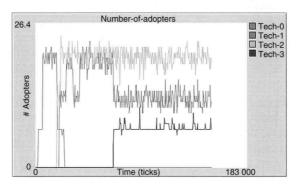

Figure 6.3   An example simulation run, with four technologies and 30 agents, showing the impact of newly invented technologies on the adoption of existing ones.

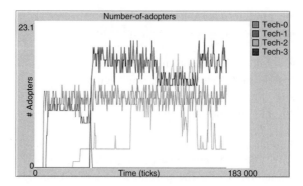

*Figure 6.4   A second simulation run with the same constraint set as before
but different random number streams governing agent actions
and movement.*

roam up to four units from their home bases, and may imitate anyone
within a radius of two units from their current position. All technologies
are simple (*Size-Of-Complex-Techs* = 1) and the chance of adaptation
during adoption is set to 0. The statistics are output every 400 time steps.
According to the figure, tech-1 was invented almost immediately and
taken up by about 20 agents. The appearance of tech-2 and then tech-0,
however, undermined the value of tech-1 completely, and it disappeared
from the population. Tech-2 reached the same level of popularity, as did
tech-0 for a period, though the invention of tech-3 seems to coincide with
tech-0 stabilising around 12 adopters. So, this example shows how the
introduction of one technology can act as a catalyst for the take-off or
demise of another.

Another important lesson from the model is that the order in which
technologies appear can matter to the long-term outcome. Figure 6.4
shows a second simulation run, using the same stream of random numbers
to generate the same constraint sets as in the previous figure, but then
using different random number streams for the agents movements and
actions. In this second run the success of tech-1 is again short lived. But
interestingly, tech-3 is able to obtain the best position, despite some competition from tech-2 which this time is relegated to a much smaller share
of the population.

The screenshot in Figure 6.5 shows the population for the first of these
two simulation runs. The agents' labels reveal their current technology
portfolios. It can be seen that these portfolios vary mostly with terrain.
Those agents in the 'grasslands' (top left) have adopted tech-2 and tech-3
(signified by '0 0 1 1'). Those in the 'forest' (top right) have adopted tech-0

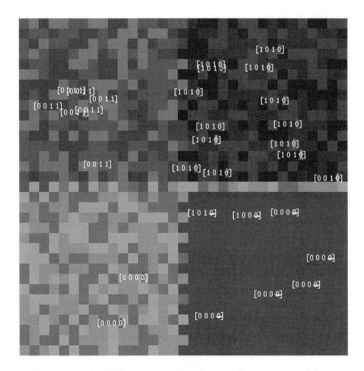

Figure 6.5   Agents in different terrains favour different combinations of technologies.

and tech-2 ('1 0 1 0'). Those in the water (bottom right) have mostly failed to adopt anything, despite the presence of 'boundary spanners', that is, agents crossing from different terrains bringing with them technologies to be imitated.

The possibilities of catalysis and path dependence can be seen using just basic technologies. Turning now to more complex technologies, we set the total number of basic technologies to eight, and divided these into two complex technologies of length four. Now that the processes of invention, imitation and discarding operate on complex technologies, a process of adaptation of the complex technology can be simulated: a removal or addition of one of the basic technologies making up the complex one. Figure 6.6 shows that adapting while imitating leads to greater average satisfaction, where an agent's satisfaction is the number of constraints (out of 16) its technology portfolio currently satisfies. One hundred per cent adaptation, of course, would be too much; nothing would ever get reliably communicated and no one would learn from anyone else. But the advantage of a fairly high adaptation rate is clear enough. A similar pattern

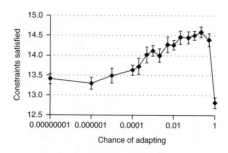

Figure 6.6   *Average agent satisfaction against chance of adapting during imitation.*

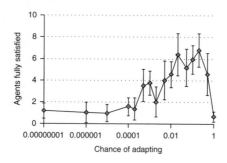

Figure 6.7   *Number of fully satisfied agents against chance of adapting during imitation.*

applies when the number of fully satisfied agents is examined (Figure 6.7), though the variance in the statistics is greater. (Both charts show aggregate results from 40 simulation repetitions.) Testing complex technologies of length 8 and 2 showed similar results. Too little adaptation leaves agents dissatisfied.

While agents become more satisfied through adapting complex technologies, the technologies themselves can benefit from higher adoption rates – or some of their basic components can, at least. Figure 6.8 plots the number of adopters of the most popular basic technology. The rate of adaptation here causes a difference between 17 adopters and 26 adopters.

In conclusion, then, the constraints between technologies produce complex adoption behaviour, including variable outcomes and path dependence. But on average, allowing adaptation of the diffusing innovations leads to greater user satisfaction and higher adoption of the most successful basic components.

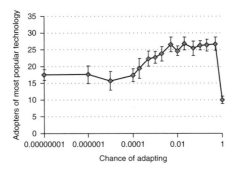

Figure 6.8   *Number of adopters of the most popular basic technology,*
against chance of adapting during imitation.

DISCUSSION

The agent-based model combines several aspects of previous diffusion models. There are social interactions, as in the epidemic model, here occurring in a geographical space. Like the probit model heterogeneous agents make adoption decisions in a dynamic environment. Like the stock and evolutionary models, technologies affect each other's diffusion and the system is capable of path-dependent outcomes. The simulation model makes visually apparent the importance of context in determining the success of an innovation. Variation in environment, represented here by 'terrains', causes variation in the value or function of a technology. In addition, for a particular technology variation in the other technologies held in an agent's portfolio, its technological context, also causes variation in the value or function. In turn this leads to variation in the success, or amount of take up, of the technology, including how far it diffuses. Once one considers complex technologies, made up of multiple basic components, allowing adaptation of the complex technology leads to greater adoptive success. A complex technology in one terrain or context may need adaptation to flourish elsewhere. This adaptation is more likely to come from users than be intended by the original innovator, though products can sometimes be designed to restrict or facilitate user modification. Diffusion success should perhaps be recognised more often as stemming from a collective effort, rather than one original individual inventor or developer. For many innovations, however, ownership of the design, credit and financial rewards tend to be directed towards a single point of origin. Adaptation by others challenges this practice. Meanwhile lack of adaptation threatens to limit adoption. A simulation of adopting and adapting illustrates why this tension occurs.

Any lessons drawn from the simulation should be tested for sensitivity to parameter settings. These include the frequencies of invention, imitation and discarding actions, the movement behaviour, the roaming and imitation radii, and the population size and population density, controlled via the dimensions of the simulated world, as well as the numbers of terrains, technologies and constraints, and complex technology size.

Although more complex than traditional diffusion simulations, the model that communicated these points is still relatively simple. A number of avenues exist for enhancements, however. For convenience this chapter considered only sets of constraints for terrains. One extension would be to add constraints for different types of agent, where agent types, unlike agents' technology portfolios, remain fixed during the simulation. Another extension would be constraints between technologies held by neighbouring agents. These could represent social and political constraints, like those identified in the above case studies in technology development projects. Agents' actions could be made less automatic. At present, agents complete an invention or imitation action if and only if it does not lower their satisfaction. A softer decision rule would be to allow agents to sometimes accept reductions in satisfaction. Such a method is found in the meta-heuristic, simulated annealing (Kirkpatrick et al., 1983). This might then enable agents to discover superior combinations of technologies later, when a further technology was found that repaired the broken constraints, or compensated for them by repairing others. Of course, for reasons of both personality and emotions people vary in their willingness to experiment with potentially costly technologies, and more psychologically realistic representations of human agents could be modelled.

The results described in this paper suggest that a visual model which combines diffusion and constraint satisfaction can be used for teaching these concepts as well as those of interdependencies, context dependence and path dependence. Simulation tools have been used to facilitate group discussions and promote understanding (Robinson, 2001). As visual tools they provide non-technical summaries and feedback to groups audiences. As simulations of complex adaptive systems they help one come to terms with the emergence of often surprising outcomes from systems of many interdependent components. In the case of innovations, this could include facilitating understanding of why some product failed to take off after launch, diffused among people in one place but not another, or went into decline after enjoying stable support for a period.

A key issue for such use of a model is how to validate it, or more generally how to obtain 'buy in' or support for its conclusions from stakeholders and clients. For traditional models this appeared relatively easy: one collected data on the number of adopters over time, and then demonstrated for some choice of parameter values how well the model's behaviour fit the data. As we discussed in Chapter 2, even then the model might be of little use in forecasting. If most of the innovation's diffusion has yet to occur, fitting what is only the start of the S-curve to data may give little information about the peak adoption rate or the saturation level of adoption. Such forecasts become much more accurate as data collection nears the peak rate, but by then there may be little time left to act on the forecast.

It has been the argument of this chapter, however, that technology innovation and diffusion often occurs under more complex conditions than those modelled with traditional approaches. The warrant for arguing so comes from ethnographic studies, available in the literature on actor–network theory and the social construction of technology. Such studies are not numerous enough to employ the methods of quantitative researchers. Indeed, their results imply that it is problematic to claim identity for the objects of two studies made at different times or by different researchers – the essential assumption behind time-series measurements and classical statistics random sampling. Predictive models are not unheard of when dealing with complex social systems, however. Axelrod (1997a, Chapter 4) simulated political alliance formation among 17 European countries during the 1930s, essentially as a problem in weighted constraint satisfaction. Not only was the most common end result from the model an alliance very close to the actual historical outcome, but there was only one other end result. Thus Axelrod's model not only helps one understand why history occurred as it did, but also provides warrant for a counterfactual history: what if events had been slightly different? What other outcomes might plausibly have occurred? In the language of complex adaptive systems, there are just two attractors in this system's state space, or two peaks to climb on a fitness landscape. Other systems might have many more such attractor states, or peaks, and finding the best becomes a hard problem, as implied by the studies of K-Sat problems (Gomes and Selman, 2005; Kirkpatrick and Selman, 1994). In our diffusion model this prompts the questions, could there have been alternative technological developments and diffusion successes to those that occurred, and if so how many?

It remains to be seen whether a case study in technology can generate a sufficiently detailed list of relevant actors and constraints to produce analogous conclusions. In the case of the development of the bicycle, Bijker (1995, p. 52) suggests that drawing up a network diagram of products, users, problems and solutions would be infeasible, due to the 'immense complexity' involved. Law and Callon (1992) provide a partial network for the TSR.2. Latour (1996) leaves the reader to make his or her own mind up about 'who killed *Aramis*', and about whether the novel transport system could ever have succeeded. The main contribution of all three accounts is delivered through their narrative, rather than a rigorous demonstration of logical incoherence between statements. It may be that diffusion models should be considered in the same way, with the model as a tool for eliciting, during group discussion, lists of components, interested parties, their interests and the interdependencies between all of these. When we understand how the meaning of an innovation varies with its context, and how this variation increases costs and project failures, it becomes easier to accept the notion of allowing adaptation to local and user demands. The various diffusion models focus attention on different aspects, and encourage different managerial initiatives. The conclusion here is that there is a role for more diffusion models focusing on complexity.

Adoption decisions vary with complex, local conditions: the decision maker's personal characteristics, social and physical environment, and what else they have adopted. Qualitative research can alert us to the complexity, but it does so retrospectively. We can understand something of why people acted as they did, when they did. What we would like to be able to do is predict how they will act in the future and why, so that we can do something about it. But our response to a prediction about a situation may be actions that disturb the situation in such a way that the predicted outcome is now obstructed. Before that, the act of gathering data, from which to make a prediction, may already have initiated changes so that the data are out of date by the time we have analysed them. And the complex interdependencies may mean that whenever our data omit details, those omissions can turn out to have crucial consequences for our attempts to understand a situation. If adoption decisions occur in complex environments, it seems implausible that we may gather sufficient market data to be able to design or select an innovative product that will then be adopted widely across varied environments. Instead of trying to predict what will satisfy a particular adopter, a better innovation policy might be to develop adaptable innovations, products that can readily be adjusted to local needs. Instead of trying to produce clear statements of their ideas, which may then be judged clearly wrong by those outside the small circle we operate in, innovators might do better to promote vaguer, more ambiguous ideas that outsiders can then reinterpret to their own ends, with benefits for both supplier and recipient. Too much ambiguity can be a bad thing, of course, as then nobody succeeds in communicating anything to anyone. But innovative solutions to problems in complex environments are more likely to be generated through the use of *abductive* reasoning, which generates plausible inferences from the juxtaposition of thoughts not directly connected to each other, analogy and metaphor, and storytelling, rather than through data gathering and analysis, or insistence on clear statements of facts. Knowledge-management guru

Dave Snowden's Cynefin framework for policy making may provide a guide here (Kurtz and Snowden, 2003; Snowden and Stanbridge, 2004; Snowden and Boone, 2007). There are at least four different environments in which we can find ourselves. In the domain of *the known* quantitative data gathering and statistical analysis thrive. In the domain of *the knowable*, system parts are interdependent but few enough in number for them to be mapped and their dynamics modelled. In the domain of *the complex*, greater interdependencies prevent prediction, but patterns may emerge which can be identified and reflected on retrospectively, and for which agent-based modelling seeks to explain by postulating and testing generative mechanisms. Finally, there is the domain of *chaos*, or disorder, for which no prediction or explanation is possible, and the best approach is to act with the hope that one's actions serve to push the system into one of the other domains. The diffusion of innovations has often been taught as if we were only ever in the domain of the known, with occasional extensions to the domain of the knowable. Innovation adoption, however, occurs more often in a complex domain, requiring flexible, adaptable interventions, and diffusion models should reflect that.

NOTES

1. This chapter is based on a paper given at the DIME Final Conference 2011, University of Maastricht. The argument was presented earlier at the EASST010 conference, University of Trento, and at a research seminar in the Centre for Research in Social Simulation, University of Surrey. We thank all audiences for their comments.
2. The program includes several alternative values for the parameter *Agent-Movement-Method* besides the 'Around base' method described here. The reader may consult the program for more information on these. An interesting question would be what difference to diffusion outcomes, if any, does the choice of movement method make.

7. Technological evolution and innovation networks

INTRODUCTION

This chapter examines models of technology and our relationship to it. By technology here is meant the collection of individual technologies, practices and ideas. All of these are themselves the result of combinations of preceding technologies, and some of them may form part of future technologies. By our relationship to technology, we mean the ways in which technologies impact on social organisation, including what we do, who we do it with and where we do it, but also the ways in which social organisation impacts on technology, especially through the invention, development and diffusion of new technologies. While previous chapters have covered the diffusion (Chapters 2, 3 and 6), the generation (4, 5, 6) and the impact of innovation (3, 4, 5, 6), this chapter describes moves towards incorporating all three processes in single models.

In Chapter 4 on models of organisational learning the idea was introduced of innovation as a search for novel combinations of organisational routines. Search practices employed in organisations included trial-and-error experimentation and learning from others, but some combinations of search practices performed this search better than others, especially where social network structure and the relative rates of learning were concerned. In Chapter 3 on social network structure we saw some of the network phenomena that might explain this. In Chapter 6 on adoption as adaptation we moved from searches for good combinations of routines to combinations of basic technologies. Interdependencies between these meant that the adoption of some could catalyse or inhibit the adoption of others, and the order in which technologies appeared mattered. Path dependence meant that if technologies appeared in a different order, they could experience different success rates. The idea of innovation effort as search is now extended beyond organisational routines and technologies to social organisation. In this chapter we see how network structures are also the outcome of innovation.

This raises the possibility of a cycle: innovation in the form of technological evolution or knowledge dynamics influences the emergence of networks of innovators, and these innovators in turn influence further evolution of technology and knowledge. This chapter will describe several models of technological evolution, knowledge dynamics and innovation networks, including some that complete the cycle.

In doing so, we will identify two issues for would-be modellers of real-world socio-technical systems. The first issue is that in each model there is a core component that is extrinsic to the model and must be specified by the modeller, yet its structural properties have a key influence over the results of innovation efforts, and hence over the behaviour of the model. In the models of organisational learning this component was the fitness landscape. In the model of innovation as constraint satisfaction this component was the lists of technological constraints. In Chapter 5 on science models we discussed the possibility of deriving a knowledge structure from empirical data, in this case data on academic publications. In this chapter we mention some more knowledge structures that connect technologies and units of knowledge.

The second issue for modelling real-world socio-technical systems is that the people and firms making them up have material properties, just as the technological objects they wield do. Technological innovation alters social organisation, including the formation of innovation networks. But the altered social organisation may have different resource demands, or incur new constraints on its operation, such as traffic problems. The changes in social organisation thus create new demands for technological solutions to problems that were not faced previously. When the technological solutions are discovered, however, this can have an impact on the social organisation as well as on the use of other technologies. Some new technologies have led to the emergence of new ways of living, and novel social structures. Through the emergence of novel technologies and social structures the socio-technological system grows over time. None of the models we survey in this chapter manage to represent this growth. But they will help us to understand better what is missing, and we will conclude with some pointers towards a future generation of models of technological evolution. How policy makers and managerial decision takers can learn from such systems will also be discussed.

TECHNOLOGICAL EVOLUTION AS CRITICALITY: SILVERBERG AND VERSPAGEN'S PERCOLATION MODELS

Silverberg and Verspagen (2005, 2007) have proposed models of technological evolution that attempt to explain patterns in the occurrences of innovations of particular sizes. Two basic ideas underlie the mechanisms in their models: that the appearances of particular new technologies are dependent on previous technologies, and that the discovery of new technologies is a process of search among designs close to what is already known.

The first idea combines two facts: that new technologies are composed of existing technologies as components, and that the value of particular technologies rests upon the existence of supporting technologies, a point we have already made when discussing the stock model of diffusion (Chapter 2) and the model of innovation adoption as constraint satisfaction (Chapter 6). A classic illustration is provided by the automobile. Motorcars consist of a variety of different sub-technologies: the petrol-driven internal combustion engine, pneumatic tyres, suspension and so on. In some cases alternative component technologies could be provided, solid rubber tyres instead of pneumatic ones, for example, or diesel combustion instead of petrol, but each of these is providing a function that the combined whole could not work without. The value of a car depends also on supporting technologies, such as relatively smooth tarmac roads, the provision of petrol stations, car breakdown services, spare parts suppliers, road repair, driving schools and so on. What we have, in fact, is a web of inter-related technologies. As cars, trucks and motor-driven buses became more available and useful, another web of technologies declined. Not only did the horse and cart become obsolete, the markets for carters, blacksmiths, farriers and haymakers also declined – the innovation of the automobile led to Schumpeter's 'perennial gale of creative destruction' (Schumpeter, 1943, p. 83) blowing away this web of economic relations. The change did not occur overnight – horsepower was the most popular technology for transport long after the workshop of Benz attached a motor engine to a cart – but viewed with sufficient passage of time, a major technological shift can be discerned.

The second idea behind the models of technological evolution, that innovation involves a search process, can be traced back in the evolutionary economics literature to Nelson and Winter (1982) and before that to Simon and colleagues on problem solving in organisations (Cyert et al., 1964; March and Simon, 1958), as was explained in Chapter 1, and drawn upon for the models of organisational learning in Chapter 4. The science models in Chapter 5 contained the same idea, namely, that novelties are sought as slight variations on what is known already, or 'local search'. However, Silverberg and Verspagen's models of technological innovation make an important distinction between discoveries, which are the immediate results of search, and innovations, which are technologies that have not only been discovered, but have also become viable. One way to think about this distinction might be to consider the difference between what can be patented on the one hand, and on the other what can be produced and used. An idea that has been patented may prove to be technologically infeasible, or may lack a market once produced. To say a technology is viable is say that a chain of supporting technologies exists, be they components, production techniques or support services, leading up to its widespread, affordable use. Clearly this is the commercially more interesting conception of innovation, since patents cost money to register, maintain and enforce but offer no guarantee of a return on investment.

This distinction is present in the model described in Silverberg and Verspagen (2005), and reproduced by us (see *SVPercolationModel.nlogo* on the website). Technologies are represented as being in one of four states:

0   Impossible (coloured black), technologically excluded by nature
1   Possible (white or grey), but not yet discovered
2   Discovered (yellow), but not yet viable
3   Viable (pink or green)

Another concept core to this model is that some technologies support, or make useful, other technologies, while other technologies can supersede or make obsolete what has come before. To represent these interdependencies visually, Silverberg and Verspagen arrange technologies in a regular two-dimensional grid space (Figure 7.1), with each technology above the bottom row having four neighbours (i.e. von Neumann architecture). In this 'technology space' columns represent different niches or functions, while the height of a row represents some degree of progression or superiority. If two technologies are in the same column, they serve the same niche or function, but the higher of the two can supersede the lower. The simulation model is initialised with all technologies being either Possible or Impossible, except for a baseline of technologies at the bottom of the space in row 0, which is set to Viable. The chance of each technology being initialised as Possible is a parameter of the model, q, also known as the percolation probability. It is known from studies of random percolation processes on lattices that a critical value of this parameter exists at which the lattice undergoes a phase shift from mostly consisting of nodes that are unconnected to each other, to mostly consisting of nodes that are connected to each other (Albert and Barabási, 2002, p. 60).

During a simulation run, technologies that are impossible remain so indefinitely. Technologies that are Possible may become Discovered as a result of R&D effort, or local search. Discovered technologies become Viable if and only if there exists an unbroken chain of viable technolo-

(a) Time = 0   *(b) Time = 1500*   *(c) Time = 2700*

Figure 7.1   Technology space showing evolution over time, with impossible
technologies (black), undiscovered possible technologies
(white), and viable technologies (shaded) developed out from
the bottom-row base.

gies linking them to the baseline. This chain may cross between columns and thus its length may differ from the height above the baseline of the endpoint technology, and also differ from a chain of technologies that are merely *Possible*. Some technologies may be initialised as Possible, but be part of a block that is cut off by Impossible technologies from being reached by any viable chain from the baseline. In our version of the model we colour these cut-off technologies grey.

R&D search is conducted each time step, for each niche (column). Search effort is expended evenly on all technologies up to and including a given number of degrees of separation, m, from the starting point of the search. If E is the total R&D effort available for a search from a particular starting point, then Silverberg and Verspagen define the chance of a particular *Possible* technology being discovered during that search by:

$$p = E/2m\,(m + 1)$$

The starting point for search in each niche is the highest *Viable* technology in that niche. This represents the state-of-the-art or best practice in that niche, and the collection of such points for all columns is called the *Best-Practice Frontier* (BPF). We colour the BPF technologies green to distinguish them from other viable technologies. When a new discovery results in some Discovered technologies becoming *Viable*, it may be that some of the newly viable technologies represent best practices in their respective niches, in which case the BPF is advanced. Previous members of the BPF are now superseded in those niches. An innovation is defined in this model as an advance of the BPF in a particular niche, and the number of rows

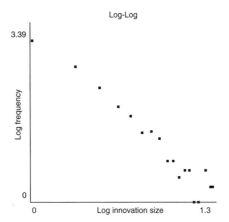

Figure 7.2   Frequency distribution of innovation sizes, that is, BPF row jumps.

the BPF has advanced in that niche represents the size of the innovation. Innovation sizes are recorded and a frequency distribution updated periodically. The distribution is highly skewed, with the possibility of some very high values. When drawn with logarithmic scales on both axes, this distribution can come to look quite straight (Figure 7.2). As mentioned in Chapter 5, when we drew frequency distributions from bibliometric data, a straight line in a log–log plot indicates a power-law or scale-free distribution. In the context of the current model it would indicate that innovations were occurring on all scales. Scale-free distributions are also encountered in models of critical systems, such as the forest fire model (Drossel and Schwabl, 1992, 1993) which might be likened to a model of the diffusion of an innovation whose appeal can pass in and out of fashion, and Per Bak's sandpiles model in which a system organises itself towards a critical state ('self-organised criticality') (Bak, 1997; Bak, Tang and Wiesenfeld, 1987; see the website for our reproductions *RepeatedForestFire.nlogo* and *Sandpiles.nlogo*). What Silverberg and Verspagen's percolation model does is turn the products of search into a critical phenomenon, a system poised between two phases, in this case between having many disconnected components and having one completely connected component.

That the sizes of innovations follow a scale-free distribution is an idea with some empirical support. Histories and ethnographies of the emergence of particular technologies reveal no eureka moment when one can clearly say a technology has been introduced. Rather, the components of a technology, the bicycle for example, appear slowly at different points in time, and are often developed for very different technological purposes than the technology whose history is being traced (Bijker, 1995). Turning to large-scale, quantitative data, Trajtenberg (1990) identified that patent data tended towards scale-free distributions of citations per patent, and estimates of the economic returns to innovations also indicate a highly skewed, heavy-tailed distribution (Harhoff et al., 1999; Scherer, 2000).

This is perhaps surprising. Various authors have distinguished between types of innovations, including 'radical' and 'incremental' (Abernathy, 1978), 'radical' and 'conservative' (Abernathy and Clark, 1985), 'breakthrough'/'discontinuous' and 'incremental' (Tushman and Anderson, 1986), and 'disruptive' and 'sustaining' (Christensen, 1997), though the basis of the distinction can vary. Possible bases for it include the scale of impact on other technologies and services, the scale of the impact on the market's incumbent firms, and the degree of uncertainty created within organisations by the appearance of the innovation (Rogers, 2003, p. 426), which can stem from its degree of dissimilarity to preceding technologies. In all these cases, it is assumed that one type of innovation, usually called 'incremental', is more common than the other type. Likewise in the history of science, Kuhn claimed to have identified periods of incremental, 'normal science', punctuated by scientific revolutions (Kuhn, 1962). In contrast, the recent empirical studies point towards there being innovations on all scales, with no basis for a distinction into two types.

Silverberg and Verspagen (2005) identified a number of other features of the basic percolation model besides the distribution of innovation sizes. First, innovations in different niches often occur together. This can be understood from the fact that clusters of discoveries can build up that are connected to each other but not yet connected to viable technologies. Eventually, a single new discovery may be made that bridges the gap to the built-up cluster, with the result that the BPF advances in multiple columns at once. Second, the amount of innovation or technological progress, represented by the mean height of the BPF, that is made during a given number of simulation iterations, increases strongly with the percolation probability, q, as more Possible technologies mean easier progress and shorter chains to the BPF (Figure 7.3). It also increases weakly with the search radius, m, though the improvement diminishes with m, and there is a curious trough around $m = 2$, for relatively high values of q, separating myopic search from longer distance search. This trough results from the fact that increasing search radii involves spreading search effort more thinly. Initially, this is a disadvantage – hence the drop in performance. But larger search radii have an advantage when it comes to avoiding deadlocks, that is, running out of *Possible* technologies to discover that can be connected to the baseline. At sub-critical values of the percolation probability, deadlocks ensue whatever the search radius. But myopic search can

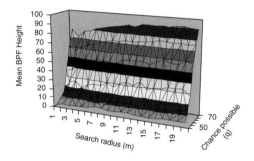

Figure 7.3 Mean height of the Best Practice Frontier (BPF) after 10
simulation runs, for various values of search radius (m) and
the chance of a technology being Possible (q).

suffer from deadlock at higher values of q. At these values there may still
exist dead ends. If a chain advances the BPF into one of these dead ends,
a large search radius can still lead to discoveries outside the dead end, enabling the blocking *Impossible* states to be worked past. But myopic search
may progress into the dead end too quickly, leaving it unable to continue
construction on other chains.

The percolation model is a visually appealing demonstration and
easy to explain, but how good an analogy with technological evolution
has it provided? The grid network provides interdependencies between
technologies. Developments in one niche can also benefit from those in
neighbouring niches. Innovations are cumulative – they build upon what
has appeared before – and recent ones make previous ones redundant.
This contributes to the generation of the frequency distribution of innovation sizes. But the model has not provided for the possibility of qualitative
innovation. The number of technological niches is fixed at the start, as are
the neighbourhood relations between niches. Within a niche, quantitative
jumps can be made of any size, including, occasionally, some relatively
large advances. But new niches cannot appear. Are all innovations really
just quantitative improvements in existing functions?

Turning to the quantitative innovation, it is not immediately clear that
the frequency distribution of BPF advances bears much analogy to the
empirically based scale-free distributions associated with innovation,
namely citations per patent and economic returns. Even if we accept BPF
advances as a good proxy for innovations, we need to evaluate the goodness of fit of a power law to the frequency distribution – i.e. how well the
straight line on a log–log plot fits the data points – and to consider what
is implied by deviations, such as when the highest-frequency, smallest-size

point falls below the trend, as in Figure 7.2. Comparisons with the scale-free distributions found in the sandpiles and forest fire models (Bak, 1997; Bak et al., 1987; Drossel and Schwabl, 1992, 1993) and in the bibliometric data (Chapter 5) may also be informative.

Real technological evolution is the result of activities by many different actors, who are each capable of complicated and dynamic behaviour. However, the model of Silverberg and Verspagen (2005) lacks multiple firms, and lacks an economic side to innovation – returns to investment in R&D are not modelled. Innovations are not used for any purpose other than providing a new starting point for search and a viable bridge to more advanced technologies. A revised and extended percolation model by Silverberg and Verspagen (2007) begins to correct for this. Searches are conducted by firms located in particular technological niches, with the option for firms of changing to nearby niches if superior progress has been made there. This enables firms to track areas of progress, leading to clustering of firms in particular regions of the technology space, to concentration of R&D search in these regions, and hence to further progress. For a wide range of parameter values it also seems to improve the approximation to a scale-free distribution of innovation sizes. Silverberg and Verspagen suggest that by making the search agents adaptive, the system is now an example of *self-organised* criticality.

Another feature of Silverberg and Verspagen's 2007 model is that firms' R&D effort is supplemented by a bonus based on any innovations that occurred as a direct result of discoveries made by that firm in the previous time step. However, since in this 2007 version of the model only one firm per time step performs search, the chance of the same firm being chosen to perform search twice in a row seems slight, and hence it is not clear that firms currently benefit from this bonus resourcing in the results reported in the paper. With some modification, however, a model could represent a process of cumulative advantage, with success in innovation leading to more search, and hence more success in innovation. Firms with inadequate success might even be allowed to go bankrupt and cease searching.

EVOLVING TECHNOLOGIES

The percolation models of technological evolution do not attempt to represent real technologies. The dependency relations in technology space are based on an arbitrary model, a two-dimensional regular grid. The roles played by individual technologies are purely to provide stepping-stones for further search and growth of chains of viability. No individual technology has any purpose or value in the model beyond this. One variation would be to simulate the evolution of technologies with genuine uses. Computers can certainly be used to evolve solutions to real problems: in Operational Research, heuristic search algorithms are used every day to seek good designs in the form of scheduling of jobs, allocating of resources and timetabling of transport services, and we noted in Chapters 4 and 5 how simulation models of organisational learning and science bear some analogy to these algorithms. But while these are examples of computer algorithms being used to design something, they do not emphasise the cumulative role of technology. Technologies as solutions to one design problem can become components in the solution of further design problems. In the percolation model, the cumulative nature of technology was made visually apparent. What is missing is a model containing multiple genuine design problems, each one providing a stepping-stone to others.

Genetic algorithms have been used to search for solutions to combinatorial optimisation problems, using methods inspired by the biological process of evolution by natural selection (Holland, 1975; Mitchell, 1996). A variation on these methods is to search for combinations of computer programming elements that make up well-formed computer programs capable of performing some task (Koza, 1992). Although the designer of the algorithm specifies the task, they do not specify the program to perform the task, and thus any eventual computer programs found by the algorithm can be a surprise to the algorithm's designer. This search method, known as *genetic programming* (GP) (Koza, 1992, 1994, 1996, 1999, 2003), has been used to design real technologies, such as printed-circuit boards (PCBs) (Koza et al., 1996). Experiments with GP have been able to rediscover designs that have already been obtained by unautomated, human-based methods (Koza, 2003). Since these designs had already been patented, this demonstrated the potential for GP to produce designs worth protecting commercially.

However, PCB designs, like schedules and timetables, are relatively easy to specify in the language of computer algorithms: namely discrete-valued numbers. Applying GP to the design of the kinds of technologies one can hold and operate in one's hand is more difficult. One can easily automatically generate instructions for the construction of a candidate device, such as 'Connect leg 2 of part 102234 to leg 17 of part 8824'. The physical effects of following these instructions – the actual attributes and behaviour of the device – are more difficult to simulate. A biological analogy here is the difference between the digital information encoded in DNA, our genotype, and the properties of the organism that grows under the influence of this code, our phenotype. The value of the former, its fitness, is clearly dependent on the success of the latter. Models of fitness landscapes, such as those using Kauffman's *NK* model, have tended to jump from genetic information (the N variables) to fitness values, omitting the phenotypic stage.

Based on Altenburg's (1997) biologically inspired extension to the NK model, Frenken (2001) links the N variables to phenotypic functions, which then contribute towards the determination of fitness values. But what these functions represent is unclear. If intended to represent physical properties of devices, such as size, weight and speed, then both the link to the N design decisions and to the fitness, or utility value, of the device seem unrealistic. The numbers used in an NK model are arbitrary and sampled from a uniform distribution. This means that various values of one of the N variables, representing different design decision options, have uncorrelated effects on a dependent function. The physical size of a multi-component device does not vary quite as dramatically when a designer tries different options. Indeed, it is hard to think up physical properties that do. The relationship between the physical attributes and the overall utility of the completed device is also too simple in this model. Frenken calculates fitness as a weighted sum of the physical functions. But sometimes both big-but-slow and small-but-fast can be good solutions to the same problems of survival. That is, fitness and utility need to be the result of more complex interaction between the physical attributes.

Attempts to simulate the evolution of micro-biological entities (artificial life) give some idea of what is involved in combining genotypic and phenotypic evolution in one model, and introduce a potentially very useful concept for understanding evolution dynamics, the concept of *neutral networks*. In the model of RNA folding of Fontana (2006) a string of characters represents a string of RNA, the genotype. Such strings undergo gradual genetic mutation. Each new RNA string created is folded up according to some pre-defined laws of chemistry acting on the current component characters. The laws determine which types of character can be located near to each other, and which tend to repel each other. This gives an RNA string a particular shape with particular physical properties, its phenotype, which may in turn determine its fitness, or chances of being reproduced under some selection process. Depending on the chemistry, some one-character variations on an RNA string fold to form the same shape. Phenotypically they are equivalent, and the network of these phenotypically equivalent RNA strings is called a *neutral network*. Other one-character variations on the same string can produce very different shapes when folded, and different shapes can have different fitness values when it comes to reproduction. If a genotypic state space, the network of possible combinations of genes linked by one-character variations, is made up of neutral networks then a couple of interesting results follow. First, gradual mutation processes can create diversity in RNA combinations without risking fitness. Second, the same gradual mutation process can also cause a jump from one neutral network to another, partially adjacent one. If every neutral network is partially adjacent to multiple other neutral networks, then the state space for RNA becomes explorable using blind mutation: simple one-character changes can take one from one neutral network to many others. But despite the gradual, constant mutation rate, RNA phenotype shapes, and therefore fitness values as well, will show punctuated equilibria: periods of stability punctuated by rapid shifts. A human designer considering only slight, one- or two-character variations on a current RNA string would mostly see only solutions on the same neutral network and thus miss many of the possibilities for jumping to adjacent neutral networks. That is, a human designer does little better than a blind process.

Moving back from theoretical biology to technological evolution, if technology space contains neutral networks, then this would explain why human exploration and imagination provides so little foresight to designers and creative workers. Most tinkering – trial-and-error experimentation through testing of slight variations – will leave one with designs of equivalent usefulness. Some funders might give up resourcing such tinkering as a result, since the R&D effort produced variations indifferent in value. But if permitted to tinker for long enough, the creative person can progress along more of the neutral network, until eventually tinkering takes him or her to another neutral network with radically better usefulness. Like in a chess game, foreseeing this later, significantly successful move is impossible at the start. The sizes of the neutral networks in technology and in the arts are greater than human imagination's 'look-ahead' ability.

Attempts to simulate the evolution of technological designs rather than artificial RNA shapes will require more sophisticated representations of the morphological properties, or physical shapes, of the technological objects and their intended environments. Models of language evolution may be easier. Words and phrases are technologies in the sense of tools, but their physical properties, especially their interactions, are perhaps easier to simulate than those of, say, the tools found in a mechanic's toolbox. Empirical studies of languages exist to inform such modelling (Solé, Corominas-Murtra and Fortuny, 2010; Solé et al., 2010). But surveying such models is beyond the scope of this book.

In order to illustrate useful technological evolution, a simpler type of design problem is needed, one of the easy-to-encode examples from GP. Arthur and Polak (2006) base a model of technological evolution around the design of logic gates. An electronic circuit or chip is desired that will provide a particular logical function, a translation of combinations of binary input states into binary output states. A base unit is available, a

Table 7.1   *Truth table describing the logical functions for a two-input NAND gate, negation (NOT), conjunction (AND), disjunction (OR), the material conditional ('Implies') and exclusive-OR (XOR). Combinations of NAND functions can reproduce the other functions.*

Inputs:		Outputs:					
X_1	X_2	$NAND(X_1,X_2)$	$NOT(X_1)$	AND	OR	Implies	XOR
True	T	F	F	T	T	T	F
False	T	T	T	F	T	F	T
T	F	T	F	F	T	T	T
F	F	T	T	F	F	T	F

device providing the simple logical function known as NAND. Given two input states this returns the output state TRUE if and only if it is not the case that both inputs are TRUE. The truth values in Table 7.1 define NAND and several other common logical functions. Given enough copies of this initial unit, it is possible to construct circuits that provide any other logical function. For example, the negation function, NOT, can be created by applying the same signal to both inputs of a NAND function. Negating a NAND function in this way will create an AND function. However, finding a combination of NAND functions that provides another function is not always this easy for human designers. In common with other combinatorial optimisation problems, as more complicated logic functions are sought, the search space of possible designs expands at a rate that computer-automated exhaustive searches cannot match. In the terminology of computer science, the problem is *NP-complete*. So Arthur and Polak's model is using a problem from electronics which is genuinely difficult and calls for some kind of heuristic search if progress is to be made.

At initialisation there is a list of desired logic functions, the demand the technologies must satisfy. The list includes simple two-place functions, like those in Table 7.1, but also more complicated functions, such as a function for adding two multi-bit numbers. There are also two lists of technologies. One, the primitives, contains just one technology, the two-place NAND function from which all other technologies will be constructed. Technologies will be added to the primitives if and only if they match one of the desired functions exactly, and cost less than any existing primitive that matches the same function. A second set of technologies is built up with any results from combining existing technologies that prove useful, that is, that match a desired function with greater accuracy than any previous technologies. Construction of new technologies involves sampling from these two lists of technologies, plus the states 'TRUE' and 'FALSE'. The resulting technology is scored for its matches against the desired functions, and costed based on the number of basic NAND components it contains.

Technologies can be used both to provide a desired function and as components of other technologies. When a new technology appears that, through its accuracy and cheapness, renders an existing one obsolete, the components of the replaced technology lose one use, and may now find themselves obsolete if no other uses for them remain. Arthur and Polak's simulation outputs two frequency distributions, one for the numbers of uses (as components) being made of each technology, and the other for the numbers of technologies being replaced as a result of a new technology. Both distributions tend towards being scale-free. That is, most technologies are used a few times, but a small elite form the components to many technologies. Likewise, most cases of obsolescence involve few technologies, but a few are 'avalanches', or Schumpeter's 'gale' of change, with many technologies being removed at once.

In Arthur and Polak's model, the list of desired functions must be specified at the start. Experiments with lists of randomly generated functions failed to produce the same distributions of usage and obsolescence. Indeed, there was little technological progress at all, in the sense of more complex technologies emerging during the simulation run. A key requirement seems to be that the list of desired functions should contain some structure. There must be both simple technologies, which may be found relatively quickly, and more complex technologies which take the simple ones as components. Without the existence of stepping-stones, progress to the more complex ones is unlikely to occur. This is how to *bootstrap* to technological complexity.

Without the complex technologies, the achievement of simple technologies does not lead to sustained progress. But Arthur and Polak's model still contains a limit to growth. The list of desired functions remains fixed during the simulation run. In theory, a sufficiently long run might see extremely cheap technologies emerge for all the desired functions, with no further changes appearing unless cheaper versions could be found. Will such a model continue to show scale-free-distributed gales of obsolescence? If not, could a model be conceived whose critical behaviour was more sustained?

ENDOGENOUS SOURCES OF DYNAMICS

In some models, the cause of dynamics can be more endogenous than in Arthur and Polak's, that is, change in one part is driven by changes in other parts, not by factors defined outside the model, such as a fixed, pre-defined list of desired functions or a fitness function or constraint list like those used in Chapters 4, 5 and 6. But endogenous sources of change are considered an important aspect of technology (Arthur, 2010). New technologies are constructed out of previous technologies, often to address problems created by other technologies. Thus technology as a whole is *auto-poietic* (Arthur, 2010, p. 2), or *self-producing*. The webs of technologies and services that build up to make use of a new technology such as the automobile or the personal computer are one form of technology's self-reflection.

How to simulate a system with this self-reflection and auto-poietic character to technology? One highly computerised example: in financial markets different trading strategies co-adapt, each one intended to make money in an environment determined by the others (see MacKenzie, 2011b, for examples). But details of the computer algorithms being used by traders are a little hard to come by (for obvious commercial reasons).

Instead, consider the prisoner's dilemma (PD) from game theory. Two players represent prisoners sitting in separate prison cells and being interrogated about the same crimes. Communication between the prisoners is not allowed and they have no prior agreement concerning how to behave. Should each prisoner decide to cooperate with each other and stay silent, or defect and confess? If both prisoners cooperate with each other, they will receive a moderate payoff (score it as a 3). If one cooperates but the other defects, the defector will receive a higher payoff (score = 5), while the co-operator receives the worst payoff (score = 0). If both defect, they both receive a payoff nearly as bad as that (score = 1). Although the best total players' payoff is obtained through both cooperating, each player is tempted by the possibility of a higher personal payoff if they instead defect while the other does not. A common extension of this game (the repeated prisoner's dilemma game) is for the players to play multiple rounds, each player knowing what combination of moves (cooperate–cooperate, cooperate–defect, defect–cooperate or defect–defect) were made in the previous rounds, and so having the opportunity to respond. For example, the strategy of 'tit-for-tat' involves punishing recent defectors by now defecting against them, and rewarding previous co-operators by now cooperating.

Strategies for playing this game can be thought of as the components in technologies. Each rival technology for playing the game consists of two components, the strategies belonging to the two participating players. Axelrod showed how to apply genetic algorithms to the problem of evolving strategies for the prisoner's dilemma (Axelrod, 1997a, Chapter 1). In Axelrod's original, genetic algorithms (GA) were used to test whether a rival to the 'tit-for-tat' strategy could be evolved that would score more highly against the set of strategies submitted to a prisoner's dilemma tournament he organised. The set of strategies a candidate was evaluated against, its environment, remained fixed. In a variation on this, suppose each strategy was instead evaluated by playing games against the other current members of the population of strategies. If future generations contain different strategies in different proportions, then the expected fitness of a particular strategy will change over time (Lindgren, 1992). Pairs of PD strategies are the components in a technology for producing strings of action pairs (cooperate–cooperate, cooperate–defect etc.), given the inputs of past action pairs. Under the GA, survival in the strategy population requires obtaining a limited resource, places in the next generation, which are allocated on the basis of the prisoner dilemma game score. Each strategy component obtains its score through beneficial interactions with other strategies, but its place in the next generation is obtained at the expense of other strategies, possibly those strategies with which it obtained high scores. Thus a particular strategy can be a victim of its own success. But some strategies may be able, for a while at least, to form a mutually supporting set. The component strategies are themselves complex, however, in the sense of consisting of multiple parts. For instance, a strategy for games between two players with memories of three rounds requires $(2 \times 3) + (2^{(2 \times 3)}) = 70$ bits to encode it. Under the GA's processes of variation, namely mutation and crossover, new strategies can appear constructed from these component bits. A new strategy may offer some of the other strategies higher scoring games than they currently enjoy. If its games provide it with sufficient reward, then it may earn itself a place in later generations, perhaps supplanting another strategy altogether.

Simulations of these dynamic PD populations show complex behaviour. The population tends to evolve towards sets of strategies that produce mutually cooperative behaviour. But occasionally cooperation may break down as parasitic strategies appear and grab many places at the expense of the strategies they score well against, leading to a shift towards generally uncooperative behaviour. However, by displacing the strategies that benefitted them – in effect consuming their food source into extinction, or driving their suppliers into bankruptcy – these parasites' success tends to be short lived, and as they die out, mutually cooperative strategies often re-emerge quite quickly. Depending on the rates of mutation and crossover the system behaviour can resemble punctuated equilibria – periods of relative stability in the population, punctuated by shorter periods of turmoil. Figure 7.4 was produced from a single 4000-generation run of the NetLogo program *GAPD.nlogo* (see website), with population of 80 strategy instances, mutation chance = 0.01, crossover chance = 0.1, memory of one round, 15 rounds per game and each instance of a strategy being

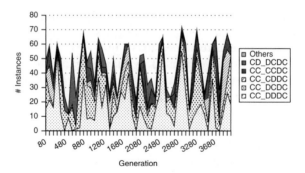

*Figure 7.4   Numbers of instances of prisoner's dilemma strategies over
time for a population of 80 instances subject to a genetic
algorithm. Frequencies were calculated every 80 generations.
The five most common strategies (which account for about 57
per cent of the total number of instances) are shown explicitly.
The most common strategy ('CC_DDDC') shows periods of
popularity punctuated by collapses in numbers.*

evaluated by playing games against 19 other strategy instances. During
the run, 61 distinct strategies appear and the distribution of their frequencies is highly skewed, with 5 strategies accounting for 57 per cent of the
total instances. The first two letters in a strategy name represent a player's
assumption for, respectively, its and its opponent's first moves. The next
four letters represent a player's response to a previous round in which,
respectively, both defected, it defected but its opponent cooperated, it
cooperated and its opponent defected, and both cooperated. The most
frequently occurring strategy, 'CC_DDDC' (overall accounting for 20 per
cent of the instances), involves initially acting as if both players cooperated in the previous round (encoded as 'CC_'), and then cooperating if
and only if both have just cooperated (encoded as 'DDDC'). This strategy, which will always cooperate against copies of itself, enjoys periods
of relative popularity, punctuated by collapses in numbers. The fortunes
of the next most frequently occurring strategy, *tit-for-tat* ('CC_DCDC'),
fluctuate as well.

The evolving set of prisoner's dilemma strategies demonstrates the
dynamics of an artificial ecology, but the population is fixed in size. Real
ecosystems can change in size, as resources, including spaces to live in,
become more or less available for the ecosystem's constituent organisms.
Likewise, the set of technologies grows, as does the population of human
beings using this set and adding to it. A challenge, then, for future models
of technological evolution is to capture both the growth of technology

and the dynamics of obsolescence, while avoiding the need to predefine which technologies will be desirable. Designs for logic gate functions and PCBs are of use to human designers, but the usefulness came from outside of the model. Prisoner dilemma strategies owed their value purely to the current state of the model – if transferred to a different environment, with different strategies to play against, their success rates changed, but with unsustainable results.

Some simulations can be both progressive and more open-ended in their outcomes. In Kauffman's (1993, 1996) model of autocatalytic sets, the programmer specifies a chemical soup of simple elements, which can then combine with each other to form larger compounds. Some elements and compounds have the ability to catalyse the formation and / or splitting of compounds, and the more compounds appear, the more likely it becomes that an *auto-catalytic set* emerges of elements and compounds, each member of which is the result of a reaction supplied and catalysed by other members of the set. *Artificial* or *Algorithmic Chemistry* (Dittrich, Ziegler and Banzhaf, 2001; Fontana, 1992) also involves simulated elements combining and affecting the outcomes of other combination events. In this case, the elements are algorithms, or mini programs, represented as strings of characters. Each program performs operations on strings, and thus the programs can transform copies of themselves and other programs. *Artificial Life* or *ALife* (Bedau et al., 2010; Langton, 1995) attempts to reproduce some of the qualitative properties of primitive biological life forms by evolving simulated life forms in an artificial environment.

Both Kauffman's and Fontana's models are *progressive* in the sense that, when supplied with basic component materials, more complex entities emerge. They are also *open-ended*, in that which complex entities will emerge during a particular simulation run is neither specified by the programmer nor can be reliably forecast. Perhaps future models of innovation can learn from these simulation approaches. Recall from Chapter 4 how organisations were conceived of as consisting of human agents following routine practices. The models of organisational learning distinguished the combinations of 'routines' being sought from the meta-routines being followed during heuristic searching. But in the artificial chemistry of Dittrich and Banzhaf (1998), the bits of information being processed are themselves routines (executed by a computer) for generating new lists of routines. In an analogous manner, the 'routines' represented in models of organisational learning could be interpreted within the simulation itself, and have meanings such as determining how much trial-and-error experimentation and learning from others is performed by a particular member of the organisation. Routines for recruiting new members and training them in the organisation's routines could ensure

the sustainability of the organisation. But the interactions of organisations require time, venues and participants, all of which may be in limited
supply. So a simulation should include resource competition. If routines
that affect other organisations can vary in which organisations they affect
most, and in when they affect them, then the possibility exists for groups
of organisations to be mutually supportive or antagonistic, like the prisoner dilemma strategies.

Such a simulation of collective search might include a broader range
of search methods than is usually allowed for. By contrast, Arthur and
Polak's (2006) model included just one algorithm for creating new technologies and deciding which if any should be replaced, generating just one
path of system evolution, with just one set of technologies. Technology
replacement, when it came, was instant and exhaustive. Real technological
evolution, however, involves multiple, simultaneous invention attempts
being performed in parallel, slow diffusion of the results through populations of users, and much adaptation or re-interpretation by those users.
As we have seen in previous chapters, this makes important the particular
structure innovators and adopters are organised in. Routine practices that
altered this structure would also become important. The next sections
describe models in which the structure between the innovators is both an
input and an output of the simulation.

EMERGENT INNOVATION NETWORKS

Cowan, Jonard and Zimmermann (2007) have described an agent-based
model of knowledge production through bilateral collaboration. One
hundred firms seek partners to collaborate with to attempt innovation,
and each firm's choice to collaborate with a particular other is influenced
by both the two firms' most recent experience of collaboration together
(using their 'relational credit'), and also the recent experiences of each
other held by any third parties they may have positive experience with
(using 'structural credit'). The resulting collaboration relations form
networks, and the structures of these networks can be analysed using
the usual social network analysis metrics and maps. (For definitions, see
Wasserman and Faust, 1994; also, see Chapter 3). Cowan et al. use their
model to demonstrate two major factors in determining the emergent
structure of these innovation networks. First, when more emphasis is given
by firms to structural credit than to relational credit, networks become cli-
quish, as indicated by the clustering coefficient. The second factor pertains
to the structure of the innovation task that firms face. When this task
is decomposable into separate subtasks, the resulting networks become

dense. In some parts of the parameter space, Cowan et al. also identified the low degrees of separation that characterise a small-world network.

Based on the description in Cowan et al. (2007) we attempted to reproduce their model in NetLogo, with only partial success (our version, *EmergentInnovationNets.nlogo*, is on the website). We found that the parameters for knowledge decomposition and for the relational-versus-structural credit balance do indeed affect the emergent collaboration network structure. However, our peaks in network density, clustering coefficient, degree centralisation and other network metrics do not always occur in the same parts of parameter space as those in Cowan et al. In addition, our metric values are significantly different. The authors reported to us that their original version, in C++, has now been lost (a warning to all simulation modellers to keep good back-ups!). So we are unable to investigate why we cannot match their results.

Cowan et al. (2007) represent innovation using a production function familiar to economists for its *constant elasticity of substitution* (CES). Each firm's current knowledge consists of a set of quantities Q of different knowledge elements or types. During collaboration, for each knowledge element, a weighted mean of the two firms' quantities is calculated. This becomes an input to the production function. For firms i and j and knowledge element e, *input-quantity* I_e is defined in terms of quantities Q_{ie} and Q_{je} as:

$$I_e = ((1 - \theta) \times min(Q_{ie}, Q_{je})) + (\theta \times max(Q_{ie}, Q_{je}))$$

If i and j differ in their quantities for each element, the parameter θ (*theta*) determines whether the resulting set of input knowledge consists largely of the best elements (θ close to 1), or whether the weaker firm is dragging down the collective effort (θ close to 0). Cowan et al. suggest this represents the decomposability of knowledge. If a task is decomposable into subtasks, then firms may take the best solution to each subtask, or high θ. Low θ represents that knowledge is more systemic. In this case firms may seek collaboration partners whose knowledge profile is similar to their own, since neither firm can benefit from its partner's superior elements.

To calculate *created-knowledge*, K, the CES production function is given by:

$$K = A \times [(\Sigma_e(I_e^\gamma))^{(1/\gamma)}]$$

A is a normalising constant. The different inputs all contribute to a single output quantity, and the CES implies that, for a given value of γ

(*gamma*), for example 10, if a quantity of one input element is substituted for the exact same quantity of another input element, there will be some loss in output. Whenever a collaborative attempt to produce innovation is successful – where success is to be defined below – the output from this function, *created-knowledge*, is added to both firms' knowledge quantities Q_{ie} and Q_{je} in one single element chosen randomly with preference for *input-quantity*.

It might be asked at this point whether the CES production function is suitable for representing knowledge production. (We leave for others the question whether the function is suitable for representing economic production, its original use. It may be noted, however, that most technologies are composed of non-substitutable elements. A car, for example, is composed of four wheels and one engine; no quantity of extra wheels is going to make up for a shortage of engines.) Does knowledge come in continuous quantities? We can know a lot of things, and know how to do a lot of things, but these things are not always inter-changeable or equivalent to each other, so aggregating over them seems misleading. In models in Chapter 4, we used strings of binary variables to represent in each case whether some routine practice or idea was believed (score 1) or disbelieved (score 0), adopted or not adopted. Knowledge as experience does, however, come in quantities. We might say that our firms have variable amounts of experience in different areas – making wheels and making engines, for example – and during collaboration they draw upon their collective experience in each area to produce some composite product. As a result they both gain more experience, but should this extra experience be added to just one area, or should it be shared out across all the areas involved in the production? (In the next section, we will describe Padgett's model of economic production, in which gains in experience of particular processes benefit other parts of the production cycle.) In addition, should both participating firms gain the same amount of experience? Is the person with the most prior knowledge best able to recognise and incorporate new knowledge – in which case, knowledge production is a cumulative-advantage process (Chapter 5) – or are there diminishing returns to knowledge as problems get harder to solve? Progress up the slopes of fixed fitness landscapes, for example, slows as an optimal solution or peak is reached.

Another concern with Cowan et al.'s (2007) model is that it uses the created-knowledge function twice. First, it is a factor in firms' estimation of how much knowledge they would get if they were to innovate successfully. For this, it is multiplied by a probability of innovation success to form a value for expected knowledge. Second, it becomes the actual amount of knowledge created in successful innovation. That is, firms may succeed or fail in their innovation attempt, but their understanding of how much knowledge they could get if they succeed is exactly accurate. In addition, this understanding is shared by both participants in the interaction. Cowan et al. simulate collaboration attempts for each firm every time step, and they use expected knowledge to rank each firm's candidate collaboration partners. So firms are assumed to be able to evaluate every other firm accurately, even though in most cases they have not interacted before.

Putting aside our concerns about the representation of knowledge in Cowan et al. (2007), it remains to describe how relational credit and structural credit are calculated and employed in choosing collaboration partners. For every pair of firms i and j, a variable X_{ij} contains the outcome of the most recent interaction between the firms, if any has occurred. This is set to 1 if there was an interaction and it became a successful collaboration, and 0 otherwise. This outcome determines relational credit, but its value decays over time according to:

$$r_{ij} = X_{ij} \times d^{(t - tij)}$$

where d is the discount factor, a number between 0 and 1, t is the current time step, and t_{ij} is the time step of the most recent interaction, if 1 occurred, and 0 otherwise.

The idea behind structural credit is that a firm may receive information from its recent collaboration partners about their other collaboration partners, and may use this information as a recommendation to attempt interaction with these partners of partners. The structural credit between firms i and j is defined:

$$s_{ij} = \Sigma_{k \neq i, j} (r_{ik} \times r_{kj})$$

It may bring to mind the idea of social capital mentioned in Chapter 3, including whether it is better to be embedded within a clique of like-minded others, whom one could trust to be like oneself, or to bridge a gap between cliques, filling a structural hole or spanning a boundary, brokering the flow of ideas between them and generating innovations by combining the best ideas of both cliques. If structural credit is employed when choosing partners, firms may experience an increase in their embeddedness as partners of partners become their own partners, and triangles are completed. The balance between relational and structural credit is controlled using a parameter α (*alpha*) to form total credit:

$$c_{ij} = (\alpha \times r_{ij}) + ((1 - \alpha) \times s_{ij} / CG)$$

where CG is the size of the group of firms whom i consults about j, that is, the sum for all $k \neq i$ of X_{ik}, Total credit is then used to influence the choice of potential partners. Parameters exist, however, to limit this influence. For given minimum and maximum degrees of influence, w and W, the weight given to a firm j by firm i is:

$$w_{ij} = w + ((W - w) \times c_{ij})$$

This weight is used twice. First, it is used in estimating the *expected* new knowledge to be created from collaboration. Here, the weight represents a firm i's degree of belief that a collaboration with j would be successful, that is, the *subjective probability* of success.

$$E_{ij} = created-knowledge(i,j) \times w_{ij}$$

If firms i and j then go ahead with their collaboration attempt, the weight was then reused by the simulation as the *actual chance* of success, or *objective probability*. Thus if firms' past experience, whether direct relational credit or indirect structural credit, leads them to rate the chance of success as 0.5, then that chance is indeed 0.5. This perfect reliability of judgment about future interaction chances seems unrealistic and future modelling attempts might suggest something different.

At each time step, firms are allocated partners in a matching algorithm, using expected knowledge as a guide. Firms also have expected knowledge values for trying to innovate on their own, with no collaboration. If at any point a firm prefers itself to every unmatched firm, then it is removed from the set of unmatched firms. Otherwise, the pair of firms with the highest expected knowledge is matched. Then the highest of the remaining firms are matched, until all firms have been allocated a partner. This matching problem, known as *the roommate matching problem* (Irving, 1985), is guaranteed to result in a matching that is stable, that is, the end result will be a matching such that no pair of firms exist that could change to new partners for gain greater than the loss to their new partners; the would-be new partners will always say no to a change.

This method of allocating partners has an implication for how innovation outcomes are represented. If firms were to differ in their recording of the outcome, they would differ in the relational credit they give interaction. That is, it would be possible to have $r_{ij} <> r_{ji}$. Structural credit would also become asymmetric, along with weights and expectations of future collaborations. Under these circumstances, a stable matching of firms would no longer be guaranteed.

Despite this, some revision of the way firms assess each other may be

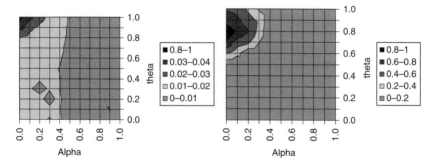

Figure 7.5   Network density (left) and clustering coefficient (right) in the emergent collaboration networks of our attempt to replicate Cowan et al. (2007).

desirable, since at present, Cowan et al. (2007) assume every firm has an accurate assessment of every other firm, kept up-to-date and ready for use every time step. Even in a marketplace with lots of publicly available data about firms, this seems a overly strong assumption.

Once firms have been allocated partners, or chosen to interact only with themselves, they attempt to create innovative knowledge. The innovation attempt succeeds with probability w_{ij}. If successful, the amount *created-knowledge* is added by both firms to one randomly chosen knowledge element.

The behaviour of our NetLogo version of the model can be seen in Figure 7.5. Varying α, the balance between relational and structural credit, and θ, representing task decomposability, the emergent collaboration networks vary in their density, clustering and degree-centralisation. Although the scales of some of our metric values contrast with the results in Cowan et al. (2007), we agree that the network clustering coefficient is highest when α is low, that is, when using structural credit – as would be expected, given that it represents a preference to close partner-of-partner relations to form triangles. We also find that relatively high θ and low α leads to a denser network – but cannot extend the relation between θ and density to other values of α, unlike Cowan et al. We tried replacing the CES knowledge production function with a representation of knowledge based on Kauffman's *NKC* fitness landscapes model (see Chapter 4). This also allows us to control the decomposability of knowledge (by varying *C*, the number of interdependencies between subtasks), but experiments produced qualitatively different results to those reported for decomposability in Cowan et al. (2007). In conclusion, then, experiments on models of emergent innovation networks seem sensitive to how we represent knowledge and innovation. This means one should be cautious before making any recommendations about real-world collaboration networks on the strength of experiments with one model. More variations on such models must be explored and reported.

The uses of relational and structural credit in models involving collaboration seems worth continuing with, however. One might expect the decomposability of tasks to interact with α, the relative strengths of relational and structural credit. If knowledge is decomposable, useful interaction partners are those whose strengths are one's weaknesses. But if firm B supplies firm A with strengths, and firm C supplies B with strengths, how likely is it that firm C will be able to supply A? C might even resemble A in its weaknesses and its use for B. In this case, we would expect structural credit to be of little use. B's high evaluation of C does not imply A will value C highly. On the other hand, if knowledge is systemic, rather than decomposable, the best guarantee of similarity in fitness values will be similarity in knowledge bits. Similarity relations show a high degree of transitivity; if B's beliefs resemble A's, and C's resemble B's, then C's will resemble A's to much the same extent. In this case, structural credit should prove more valuable than before in seeking new partners. This does not mean, however, that firms should always seek partners who are high in relational and/or structural credit. Using relational credit assumes that collaboration partnerships are worth repeating. But if two firms learn from their collaborative innovation they become more similar in knowledge. If more similar, then there is less likelihood that they will learn from each other in future. For a source of new ideas, it might seem better to avoid previous partners. The assumption in Cowan et al. (2007) that previous success is a good guide to future success seems questionable.

Nevertheless, real knowledge production involves dangers. An algorithm generating new solutions to a given optimisation problem, such as the travelling salesperson problem or a packing problem, wastes only computing time if its latest attempt results in an invalid solution. In plenty of real-world problem solving, however, unfit combinations of factors can be fatal to the agents who try them out. If solutions similar in their components resemble each other also in the danger they pose to their possessors, then for continued safety it makes sense to explore local solutions, that is, those with a high degree of similarity to what one has already found to be safe. If other firms are performing the same local searches, then firms similar to oneself, even though they offer little that is new, are still likely to provide the safest new knowledge. Firms in brokerage positions between clusters take greater risks when innovating than those surrounded by similar others.

One final avenue for future investigation is the impact of social, cultural and geographical constraints on who can collaborate with whom. Chapters 3 and 4 demonstrated how important such network relations can be when supplied by the modeller. In Cowan et al. (2007), all 100 firms are able to interact with every other. Suppose each firm was constrained in which others it could evaluate or interact with. How might extrinsic constraint network structures affect the emergent collaboration network structures? In the next section we discuss another model of emergent structures in which the existence of a prior, extrinsic structure of interaction possibilities plays an important role in making possible the emergence of collaboration structures during simulation runs.

ECONOMIC PRODUCTION SYSTEMS AS ARTIFICIAL CHEMISTRY

In a series of works John Padgett and collaborators have developed a model of co-evolution between products and social organisation (Padgett, 1997; Padgett et al., 2003; Padgett, McMahan and Zhong, 2012). Products are transformed by and transferred between firms according to some 'skills' or 'production rules' possessed by the firms. The flow of products, however, induces changes in the distribution of production rule instances and in the transfer links between firms, thus altering the possibilities for future flows. Although derived from theoretical biologists' attempts to explain the origins of life, especially Eigen and Schuster's work on hypercycles (Eigen, 1971, 1979; Eigen and Schuster, 1977, 1978a, b), Padgett's models demonstrate circumstances under which social and economic organisation can emerge from systems with no prior planning or intelligence built into them, and some relatively simple starting assumptions. The models' development represents a combination of social network analysis and Kauffman's work on autocatalytic sets (Kauffman, 1993, 1996). In Padgett and Powell (2012) it is related to empirical work on the emergence of novel forms of organisation, in particular markets, including Padgett's own classic studies of marriage, politics and business relations in renaissance Florence (McLean and Padgett, 1997; Padgett, 2001; Padgett and Ansell, 1993).

Padgett's *JAVA* version described in Padgett (1997) was bundled with earlier versions of *RePast*, and he and his collaborators are currently developing a version in *R*. Our replication in NetLogo of the model described in Padgett et al. (2003) is available on our website (*HypercyclesCW.nlogo*) and results shown here were generated using this.

There are several key components to the model: production rule instances, firms, the links between firms, and the input and output environments. Products are represented by their passage through this system, one at a time. There are also three sets of processes: the transformation and transfer of products, the learning of rule instances by doing and its counterpart, the forgetting of rule instances and, third, the passage of products to and from the surrounding environment.

The system of production rules represents a simplified version of Fontana's algorithmic chemistry (Fontana, 1992). In Padgett's model, rules relate one input product to one output product. Unlike Arthur and Polak's (2006) model of technological evolution, there is no combination of components to make compounds, nor is there decomposition of compounds. In addition, unlike Kauffman's autocatalytic relations (Kauffman, 1993, 1996), there are no catalysts raising or lowering the chances of these production rules working. There is a fixed number of types of product, n. Using numbers from '0' to represent the products, the range of rules may be written as '$0 => 1$', '$1 => 2$', ... '$(n-2) => (n-1)$' and '$(n-1) => 0$'. The last rule, of course, reproduces the first product, and taken collectively this set of rules in action represents a complete production cycle, called a 'hypercycle'. More complicated *chemistries* of rules are possible, but this chemistry, called 'SOLO' by Padgett, is the simplest one to involve all products in a single complete cycle, so it is the focus here.

At simulation initialisation, 200 instances of production rules are distributed evenly among a population of 100 firms, giving two instances per firm. Each rule instance is of a randomly chosen type, sampled uniformly from the simple 'SOLO' chemistry. Firms are arranged according to some network structure. Following Padgett we shall use a regular eight-neighbour grid (Figure 7.6), the *Moore* architecture, and compare it with a complete network, though our NetLogo program allows the user to explore other possibilities. The grid network may represent locations and proximity in geographical space. The input and output environments are common to all firms, with no distinction made for spatial location.

A production run, a chain of transformations and transfers, is simulated at each simulation time step. At the start of a production run a randomly chosen rule instance is chosen from all those among the population of firms. To be the input to this production rule, a product instance is taken from the input environment and a sequence of transformations and transfers begins. The firm currently in possession of the initial input product instance transforms it according to the rule. The new output product instance is then transferred to one of the firm's neighbours, chosen without preference. A firm receiving a transferred product instance searches its production rules for one which takes this product as an input. If the firm lacks such a rule then the product instance is discarded into the output environment and the run comes to an end. Otherwise, the firm

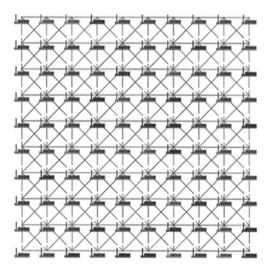

Figure 7.6   One hundred firms arranged in a regular, two-dimensional, eight-neighbour grid.

selects without preference one of the instances of compatible rules it possesses, transforms the product instance and transfers it to a neighbour as before. Providing that receiving firms continue to possess compatible rules, the sequence of receiving, transforming and transferring products instances continues until the product instance is transformed into its original type. One way to think of this is that the transformed product instance has reached its end consumer and has been transformed into money, thus enabling firms to return to the input environment for more raw materials. At this point, the production cycle is complete, the run ends and the final output product instance is put into the output environment.

The input environment is set to be either 'poor' or 'rich'. 'Poor' environments contain a single instance of the base product, '0'. 'Rich' environments contain single instances of every product. In addition, input environments may be 'fixed' or 'endogenous'. If the environment is fixed, the stocks of all product instances available are represented as indefinitely large – product instances sampled to initiate runs are immediately replaced by new instances. If the environment is endogenous, then instances removed are not automatically replaced. Instead product instances placed in the output environment are counted as additions to the input environment, from whence they may be selected to initiate new production runs. If the product instance ending the run is not of the same type as that initiating the run, then the relative frequencies of products in the endogenous input environment alter. If stocks of a particular product should eventually fall to zero, then any production runs will fail to commence if their first rule requires that product as input.

Firms 'learn by doing' whenever one firm transforms a product instance into a new type and hands it on to another firm able to transform the new type of product. Two versions of this learning are available in the program. In 'Source Reproduction' or 'egoistic learning', the transferring firm gains an extra instance of the rule it used to transform the product. In 'Target Reproduction' or 'altruist learning', the receiving firm gains the extra rule instance. Thus under the egoistic process, a firm itself benefits from having a customer who appreciates its outputs. Under the altruistic process a firm benefits from having a supplier of transformable inputs. To maintain a constant number of rule instances in the population, and to place a selective pressure on rules, each time a rule instance is replicated, a randomly chosen rule instance, which may belong to any firm and be of any type, is 'forgotten' and deleted. If this leaves its possessor firm with no more instances of any rules, then that firm exits the system. Its disappearance leaves a gap in the spatial network. Its neighbours will no longer be able to transfer products to it, and they may find they no longer have any neighbours to pass product instances to, thus leaving them unable to replicate their own rule instances through learning by doing. Firms increasing their number of instances of a particular rule have greater chance of maintaining a stock of that type for transforming future products, greater chance of one of their rules being chosen to initiate a future production run, and greater chance of employing that type of rule whenever they possess multiple rules compatible with a given product (not possible in 'SOLO' chemistry).

During simulation runs, some firms gradually lose rules and disappear. The remaining firms tend to be those enjoying instances of rules, which are particularly useful in the context of their remaining neighbours. Firms that form part of complete production cycles will tend to preserve or gain in rule instances. Likewise rules will benefit from being part of complete cycles. Firms that do not benefit from any complete cycles, through either being a part of one or receiving inputs from one, will tend to lose their rule instances and disappear. Whenever multiple cycles are realised in a population, they compete against each other by their flows. Some cycles may be realised with more rule instances than others, and some cycles may partially overlap. The product flows that occur most often tend to have the best chance of surviving, and being a part of multiple cycles may be a good way to achieve this.

The combination of production and learning bears analogy with ant algorithms and some of the cumulative advantage processes operating in academic science (Chapter 5). The more a firm and its rules are employed,

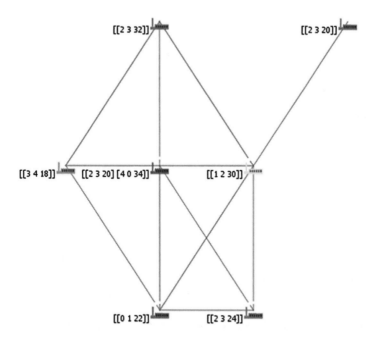

Figure 7.7   *An example system with hypercycles. Based on rule complexity of 5, this uses products from '0' to '4'. Each surviving firm is shown with a list of rules. For each rule, the input product, output product and number of instances is listed. Assuming a start with a '0' product, in this example there are then two partially overlapping hypercycles. Two firms, the top-right and bottom-right ones, are 'parasites', making no contribution to the hypercycles but receiving from them.*

the more chance they will be used in future. Over time, complex, self-supporting systems of hypercycles of rules and firms emerge (Figure 7.7), just as efficient paths to food emerge among swarms of ants. No ant can conceive of the achievement of the swarm; it just moves according to simple rules of behaviour and alters its environment as it does so by leaving a trail of pheromones. Likewise, no firm sees the entire hypercycle system it forms part of; its only dealings are with its neighbours and the environment, and learning by doing alters the probabilities with which it does that. But collectively a group of firms can find a stochastically stable organisation of firms and rules. This form of collective intelligence is the process of *stigmergic learning*, learning through the adaptation of the environment.

When the input environment is set to be endogenous and product instances output to it can become recycled as inputs to new production runs, then the model contains a second form of stigmergy. The simulation offers two options for selecting products from the input environment. Under 'Selective Search', once the initial rule has been chosen, an instance of its input product is selected from the environment, providing at least one such instance exists. If no instance exists, the production run fails. Under 'Random Search', a product instance is sampled from the environment. The relative frequencies of products in the input environment then determine the probabilities with which instances of particular products are taken out of the input environment. The production run continues only if the sampled product instance is of the type required as input for the chosen rule. Thus, which rules begin production runs depends under selective search on which products have at least one instance in the input environment, as well as on the relative frequencies of rule instances in the population. But under random search the relative frequencies of both products in the input environment and of rules determine which production run commence. Since these production runs may then alter the relative frequencies of products in an endogenous environment, the potential exists for them to alter the chances of their own re-occurrence. Under selective search from a rich, endogenous environment, stocks of some products can crash to zero. Under random search, stocks are mostly maintained, since sampling initial input products tends to correct any imbalances in relative frequencies between the stocks.

Following Padgett (Padgett et al., 2003, 2012) the model can be used to explore the circumstances under which stochastically stable hypercycles emerge from the initial, randomly generated population of firms and rules. In particular, we varied the number of products and thereby the number of rules ('Complexity'), the firms' network structure between complete and eight-neighbour grid, the learning regime between 'egoist' and 'altruist', and the input environment type between 'poor' and 'rich', and 'endogenous' and 'fixed'. Each simulation run consisted of 100 000 time steps, with one production run attempt per time step, unless the number of hypercycles in the system was detected as having dropped to zero, in which case the simulation run could be halted prematurely. Figure 7.8 shows for each parameter combination what proportion of 20 simulation runs ended in a system still containing hypercycles. In some cases, such a system might still lose its remaining hypercycles if further time steps were simulated, but in most cases the results would not have changed noticeably given simulation runs of, for example, ten times the number of time steps. The important lesson, the one of practical value, is that hypercycle stability over a particular duration varies dramatically with the parameters. Whether the

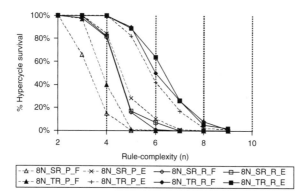

Figure 7.8   *Percentage of simulation runs ending with at least one surviving hypercycle. Results are shown for eight-neighbour grid (8N), with Source Reproduction (SR) and Target Reproduction (TR) learning, Rich (R) and Poor (P) input environments and Fixed (F) and Endogenous (E) environments. Random Search is used for initial input products. The horizontal axis measures rule complexity, i.e. the number of types of rule, which is also the number of types of product, n.*

hypercycles are stable in some 99.99 per cent of runs lasting trillions of steps seems less important in the absence of any real-world interpretation of the model.

Our results are not identical to those of Padgett et al. (2003). The results expressed in the working-paper version of Padgett et al. (2012) are not identical to those of his earlier work, either. However, in general, we are able to endorse Padgett's main points. All parameter combinations show a phase shift from 100 per cent hypercycle survival at low rule complexity to 0 per cent survival at higher complexity. Different parameters values determine over what range of rule complexity that phase shift occurs.

The first point is the role of distributing rule instances among firms in a network. It is known from analytical studies that when rule instances are sampled at random from a single location, the stability of the system depends on rule-complexity, the length of a cycle of rules (Hofbauer and Sigmund, 1988, 1998). At rule-complexity $n <= 3$, the system stocks fluctuate but always around the same levels. At $n >= 5$, the system is unstable. At $n = 4$ it is usually stable, but close to collapse. The different system behaviours can be seen in Figure 7.9, produced using our program *NonSpatialHypercycle.nlogo*, with Source Reproduction of rules. The data series are labelled according to the input product of the corresponding

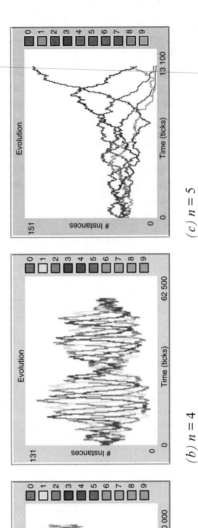  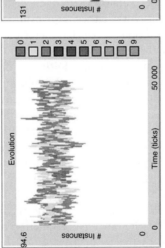

(a) n = 3 (b) n = 4 (c) n = 5

Figure 7.9   Numbers of instances of types of rule in a non-spatial hypercycle model. Behaviour over time is shown for rule complexity of n = 3, 4 and 5.

224

rule. In the unstable system, the numbers of each type of rule rise and fall in response to rises and falls in the numbers of their target rule. (Given Target Reproduction, the dynamics respond to those of the source rule.) Such behaviour may be familiar from predator–prey models, but here the peaks and troughs are growing over time. Eventually, one rule runs out of instances, and the cycle is broken forever. In this non-spatial hypercycle model, then, the divide between stable and unstable systems is strict and insurmountable. But by dividing rules among firms, the limit can be overcome. When the non-spatial model samples the next rule, it does so from the entire population of rule instances. When Padgett's multi-firm model has an output product to pass on to a new rule instance, it first samples uniformly a neighbouring firm, regardless of how many rules that firm may have. As the simulation continues, some firms will lose all their rules and disappear, while others build up stocks of particular types of rule. This separation of rule stocks from sampling probabilities enables the system to avoid the population dynamics of the non-spatial version. Network structure between the firms then influences how much the system can surpass the limit seen in the non-spatial model. Figure 7.10 shows results for a complete network. Comparing them with those for the eight-neighbour Moore-architecture grid (Figure 7.8), the complete network is able to achieve stable hypercycles at higher rule complexity, most notably with target reproduction and endogenous environments. Testing other network architectures is easy with our NetLogo version of the model. Watts (2012) examined the effects of varying network density, and compared regular, clustered networks with randomly rewired ones. The results

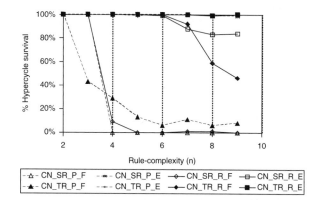

Figure 7.10   *Percentage of simulation runs ending with at least one surviving hypercycle. Results are shown for Random Search in complete networks (CN). Other labels as before.*

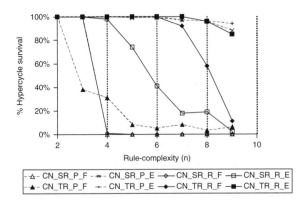

Figure 7.11 *Percentage of simulation runs ending with at least one surviving hypercycle. Results are shown for Selective Search in complete networks (CN).*

were varied, with odd values of rule complexity differing from even ones in the effects on hypercycle survival. In some cases, however, hypercycle survival chances peaked around a particular value of network density, while random rewiring, which breaks up clustering, tended to have a negative effect on hypercycle survival.

The second point to derive from Figure 7.8 is the difference between Source and Target Reproduction, or egoist and altruist learning by doing. Target reproduction is superior to Source Reproduction for all environments and both networks. Target reproduction, or altruistic learning, enables systems to self-repair. By causing the reproduction of a part of the cycle other than itself, each enacted rule maintains the cycle within which it is flourishing. As a by-product, target reproduction also maintains parasites, firms and rules that receive products from hypercycles but do not contribute to their members. Parasites are a drain on resources – they draw products away from their suppliers' hypercycles. If multiple hypercycles exist in the population, competing against each other for resources, or production runs, so as to maintain themselves, then the presence of parasites might cause their supplier cycle to lose out in competition. However, once competitor hypercycles have disappeared, the survivors can support parasites indefinitely. Future extensions to the model, such as processes that add firms, rules and links to an existing system, may even find new roles for parasites. If the parasites can acquire new target firms and rules, then they can become parts of new hypercycles. Source Reproduction allows a different kind of free-rider, firms that contribute usable products to a hypercycle (thus qualifying for egoistic learning), but do not receive products from a hypercycle. These firms could also form components of new hypercycles, given processes that added new suppliers.

A third point is that stigmergy, in the form of environment endogeneity, is most notable for its ability to improve stability chances in poor environments, whether with Source or Target Reproduction. This is no surprise. Endogeneity fixes a problem with poor environments, namely that rules with inputs other than the first product cannot become initiators of production runs. But, while in the eight-neighbour grid endogeneity makes negligible difference in rich environments, with the complete network it makes noticeable improvements in both environments.

Finally, we can compare Random Search with Selective Search. The interesting differences to be found here are in the complete networks, which were not examined by Padgett et al. (2003). Comparing Figure 7.11 with Figure 7.10 it is apparent that Random Search makes a big improvement to the chances for source learning in a rich endogenous environment. Any other differences are too small to appear without more data and more formal analysis. For the complete network, rule complexity beyond $n = 9$ products should be investigated.

Padgett's hypercycles model demonstrates that organisation can emerge without much intelligence. Given a rule, firms can search the environment for a compatible input product and, given a product, they can select a compatible transformation rule. But as Padgett puts it, this is 'the intelligence of a cow, looking for grass', in contrast to the (non-)intelligence of 'an atom, bouncing around' (Padgett et al., 2003, p. 851), and a far cry from conscious attempts to design a working organisation. Learning by doing is sufficient to lead to the emergence of organisation through the prioritising of some rules over others, some links to firms over others and some environmental products over others. There is an analogy here with how biological organisms are formed and maintained by the energy flows through their metabolic systems, and also with the maintenance of business organisations and social communities through their transactions. In particular, the organisation can transcend its particular constituent members. What matters for its survival is the maintenance of the transformations and interactions, not who is currently performing them. At the same time, Padgett's model demonstrates how the presence of a wider system shapes individuals within it, be they firms in an economy or workers in an organisation. The role played by an individual firm – its skills and its network position – are determined by its transformation activities and its transferral interactions. Padgett and Powell (2012) intend to illustrate that

in the short run, actors create relations; in the long run, relations create actors … In the short run, all objects – physical, biological or social – appear fixed,

> atomic. But in the long run, on different time scales, all objects evolve, that is,
> they emerge, transform and disappear. (Padgett and Powell, 2012, pp. 2–3)

To make its points the hypercycles model needed to be simple and abstract. To improve the analogy with real-world systems, additions need to be made to its rules, and the behaviour of its component firms. The model's 'chemistry' of rules related only one input to one output, in contrast to the SKIN model to be discussed in the next section. This omits cases where multiple components are brought together, complex materials are decomposed to form multiple products, and one product catalyses transformations of others. To simulate production rules with multiple inputs would introduce problems of synchronisation – ensuring that rules were activated if and only if their firms had instances of every input product. Firms would need to be represented with storage capacity for products awaiting the arrival of other inputs. Firms might also reflect how they are resourced. Human workers consume a far more complex range of items than those supplied to the factories they work in.

Like other models involving a heuristic search process, such as the models of organisational learning, the hypercycles model converges on a single (stochastically stable) state. Firms, links and rules can be lost from the system, while the numbers of product instances and rule instances are kept constant. More realistic firms should be able to relocate, acquire new types of skill and form social relations to new partners. An important topic is what happens when a hypercycle system, or part of one, is transferred to a new environment. Like the cross-over process in Genetic Algorithms, this seems to have a low chance of producing firm–rule–environment combinations that actually work, and in the context of the hypercycles model failure means firm and rule death. But some types of systems in some types of environment may be robust enough, and sufficiently resilient and adaptable to survive the transfer. These circumstances need to be investigated.

Padgett's own extension plans for the model include distribution rules, governing which output products are transferred to which neighbours, and the transformation and transfer of symbols rather than actual products, such as occurs in requests for input products and advertisements of newly available outputs (Padgett and Powell, 2012). Production rules may also be communicated between firms. A more advanced model might also represent its actors as having some awareness of the hypercycle systems themselves.

A key development will be the representation of different types of transformation and transfer relations. At present systems can emerge which include multiple, partially overlapping hypercycles, but the constituent rules are drawn from a single set or chemistry. Human actors can enact roles in multiple networks of relations, including socialising, business trans-actions, family, politics and religious participation, each with its own type of products – friendship, money, babies, etc. Enacting all these relations places competing demands on a person's time and energy. Fluctuations in one transaction rate may have implications for others in the same network and in networks in other domains. Padgett and Powell (2012) identify several historical examples where the coincidence of different networks appears to have played a role in the emergence of new organisational and institutional forms. Under what circumstances changes in one domain per-colate through to cause changes in others, and what determines the scale of the percolation, will be important topics for future research, especially given contemporary interest in what makes an organisation robust and resilient to change, and what makes socio-economic production sustainable.

MARKET SELECTION ON ORGANISATIONAL LEARNING

The SKIN model (Simulating Knowledge dynamics in Innovation Networks) is a model of organisational learning in which a market selects which innovation results are maintained (Ahrweiler, Pyka and Gilbert, 2004; Gilbert, Ahrweiler and Pyka, 2007; Pyka, Gilbert and Ahrweiler, 2007). (The SKIN model has its own website from whence the latest NetLogo version may be downloaded: http://cress.soc.surrey. ac.uk/SKIN/.) Like the model of Cowan et al. (2007) a population of firms engage in collaborations whose patterns can be analysed.

A firm's knowledge consists of a set of replicable units, named *kenes* (Gilbert, 1997). Each contain consists of three numbers, representing a firm's *capability* C in some scientific, technological or business domain (for example, biochemistry), an *ability* A to perform some technique within this domain, and the *expertise level* E that the firm has reached with respect to this ability. Each firm maintains a strategy, its *Innovation Hypothesis* (*IH*), for creating a new product by drawing upon a number of input kenes. This IH takes the form of a normalised sum-product of capabilities and abilities, that is:

$$P = (C_i \times A_i + C_j \times A_j + \ldots) \bmod N$$

where P is the new product, $i, j, \ldots$ are input kenes, and N is the maximum number of possible products. The new product also has a particular quality, Q, calculated as a sum-product of abilities and expertise. Obviously these definitions have no basis in reality, but they define a mathematical world of products and innovation relations in which the same product may be realised in multiple ways, and products with similarity to each other may have some similarity between their components. Variations on these definitions are clearly possible and better grounded representations of knowledge may be inserted when they become available.

A firm can only produce if it can source the input kenes. If it does not possess them itself and they are not provided by the simulation as 'raw materials', then the firm must seek them on the market where other firms advertise their products. Firms maintain prices for their products, set to cover the costs of inputs, and adjusted over time in response to supply and demand. If multiple suppliers exist for a product, a customer firm chooses the cheapest, and if more than one supplier offers the cheapest price, the customer uses the quality of the supplied input product to decide between them.

Each time a kene is used to make a product its expertise level gains 1, while unused kenes lose 1 from their expertise levels. Any kenes with expertise levels reaching 0 are 'forgotten' and removed from the set. In this respect, a process of learning by doing is operating, just as in the hypercycles model.

Firms are dependent on their knowledge base for generating sales. A number of ways for improving their knowledge bases, or learning methods, are simulated. First, if a firm no longer finds buyers for its product, it may for at a small cost undertake *incremental research*. This takes the form of an adjustment, up or down, in the ability value of one of its kenes – effectively a gradual move of the kene across knowledge space. Successful moves may be repeated, unsuccessful ones reversed, and another kene chosen if neither upward or downward adjustments are successful. A firm facing bankruptcy will take the more desperate measure of radical research. In this a kene in its IH is chosen at random and replaced to form a new IH.

A third way to improve knowledge involves forming an alliance or partnership with another firm. Firms advertise the capabilities currently in their IH. For finding a partner, two strategies are provided by the model. Under a conservative strategy a firm is attracted to others with similar capabilities. Under a progressive strategy a firm prefers partners with different capabilities. The firm also employs past experience in its search, by considering first previous partners, then suppliers, customers and finally any other firms. A partnership is offered to the first candidate whose attractiveness meets some threshold value. As a result of a partnership, the firm acquires the kenes from the partner's IH. Where a capability is new, the receiving firm adopts the kene with a reduced expertise level. Where the capability is already known, the kene takes whichever firm's expertise level is the highest, a process slightly reminiscent of collaboration on knowledge elements in Cowan et al. (2007).

Firms are given a method of increasing their production beyond the rate of one product per simulation iteration. A profitable firm may form a production *network* with recent partners, providing all members have the funds to meet the initial formation costs. This network then operates as an autonomous agent, producing in addition to its members' production, with an IH of its own based upon the kenes of the members' IHs. If it succeeds in selling products, the profits are distributed among its members. Unsuccessful networks may be dissolved.

Whenever a firm makes a particularly high profit, start-up firms may enter the market as clones. Attempting to emulate its success they copy its capabilities up to their absorptive capacity, but start with base level expertise. The population of firms can also be reduced, for example whenever a firm goes bankrupt, or forgets all its kenes through inability to use them in saleable products.

With relatively many features the SKIN model has plenty of parameters to explore. Outputs can include firm population size, number of production networks and profit over time, and the frequency distribution of sizes of production networks. This latter can be highly skewed (Pyka et al., 2007) or even approximate to scale-free (Gilbert et al., 2007). The numbers of firms and production networks can show both cycles and growth over time (Pyka et al., 2007). Gilbert et al. (2007) studies the impact on firms' longevity from different knowledge-improvement methods. At a workshop in 2011 at the University of Koblenz-Landau, Germany, projects were described applying variations of the SKIN model to innovation in the biotechnology, renewable energy, aeronautic and pharmaceutical industries. By incorporating the pricing of knowledge products and a market, it goes beyond the earlier models of organisational learning, yet, like the models of Cowan et al. (2007) and Padgett et al. (2003), it still generates emergent patterns to be analysed.

CONCLUSIONS

A range of models of technological evolution, knowledge dynamics and emergent innovation networks have now been described. To complete the survey we compare and contrast them in response to a number of questions.

Table 7.2   Representations of knowledge, technologies, strategies or rules.

Model/Source	Representation
March (1991)	Bit string. Evaluated by comparison with fixed 'environment' string. Heuristic search via trial-and-error and learning from others.
Lazer and Friedman (2007)	Bit string. Evaluated as position on Kauffman's *NK* fitness landscape. Heuristic search via trial-and-error and learning from others.
CitationAgents_NK.nlogo (Chapter 5; Watts and Gilbert, 2011)	Bit string. Evaluated using *NK* fitness. Alternative: coordinates in plane. Evaluated using Gaussian functions.
AdoptAndAdapt.nlogo (Chapter 6)	Bit string. Evaluated for satisfaction of fixed sets of terrain-based constraints (K-Sat problems).
Percolation model (Silverberg and Verspagen, 2005, 2007)	Cell locations in grid. Evaluated according to existence of link to base, and height above base. Discovery follows random search within radius of highest linked cells.
Arthur and Polak (2006)	Combinations of logic gates. Evaluated for extent to which they generate fixed set of desired logical and arithmetic functions, and for number of base components they employ.
Prisoner's Dilemma GA (Axelrod, 1997a, ch. 1; Lindgren, 1992)	Bit-string encoding of strategy for prisoner's dilemma game. Evaluated according to a fixed pay off table by simulating games against other members of variable population. Genetic algorithm updates population.
SKIN model (Ahrweiler et al., 2004)	Kenes: triples of integers representing capabilities, abilities and expertise. Combined using modulo arithmetic to generate new kenes according to an Innovation Hypothesis.
Cowan et al. (2007)	Numerical variables. Pairs of variables combined as weighted sum of minimum and maximum values to form inputs to production function assuming constant elasticity of substitution. Function output added to a one knowledge variable.
Hypercycles model (Padgett, 1997; Padgett et al., 2003; Padgett et al., 2012)	Simple production rules, or skills, for transforming one input product into one output product. Firms acquire multiple instances of rules via learning by doing. Sets of rules called 'chemistry'. Collectively self-supporting sets of rules called hypercycles.

Table 7.3   Representations of recombination.

Model	Representation
March (1991)	Organisational code updated probabilistically to consensus opinion among superior group of workers. All workers update opinions probabilistically to organisational code.
Lazer and Friedman (2007)	Agent copies part or all of solution held by neighbour with current best solution.
CitationAgents_NK.nlogo (Chapter 5) (C. Watts and Gilbert, 2011)	Co-authors selected with preference for past attempts (i.e. papers) or past successes (peer-reviewed publications). Reference material selected with preference for past usage (citations). Optional requirement for similarity.
AdoptAndAdapt.nlogo (Chapter 6)	Copying of geographically close others. Agents move within fixed radius from homebases.
Percolation model (Silverberg and Verspagen, 2005, 2007)	No interaction in 2005 model. 2007 model assigns agents to columns and allows them to change column in response to recent improvements in height.
Arthur and Polak (2006)	No social interaction. Any item in technology list may be combined with any other, with some preferential selection.
Prisoner's Dilemma GA (Axelrod, 1997a; Lindgren, 1992)	Any pair of strategies may be sampled from population for crossover.
SKIN (Ahrweiler et al., 2004)	All firms' kenes in one market. Input products chosen from market with cheapest supplier. Prices to be offered adjusted in response to demand (learning by feedback).
Cowan et al. (2007)	Partners chosen following pair-wise matching process, using on expected knowledge calculation based on past experience of interaction, reputation via neighbours, and current knowledge production function output.
Hypercycles model (Padgett, 1997; Padgett et al., 2003; Padgett et al., 2012)	Firms choose random neighbours without preference when transferring products. Passing on products useful to neighbour (learning by doing) leads to new copies of rules. Firm nodes disappear from network if all rules forgotten.

- How are knowledge units or technologies encoded?
- How are novel units generated from current ones?
- What use is made of learning-by-doing processes?
- How are the novel units evaluated? Is there progress in some sense during the simulation run?
- What patterns emerge during the simulation? Are there emergent social networks? If units are replaced, what is the distribution in replacement sizes?

The percolation model (Silverberg and Verspagen, 2005, 2007) represented technologies as cells in a regular two-dimensional grid. Technological development was represented by the gradual percolation of first 'discovered' and then, once connected by unbroken chains, 'viable' states up the grid from the base row. Innovation was defined as a rise in the best-practice frontier – the highest connected cell in each column. This provided a visual analogy to what occurs in models based on heuristic search of a fitness landscape. In particular, the models of organisational learning surveyed in Chapter 4 (Lazer and Friedman, 2007; March, 1991), and one version of our own science model, *CitationAgents_NK.nlogo*, referred to in Chapter 5 and in Watts and Gilbert (2011), all use heuristic search as a model for the advancement of knowledge or technology. Beliefs are represented as combinations of binary variables, and linked to each other by the processes of exploration, in particular, trail-and-error or mutation, and partial learning-from-others or cross-over. Unlike 'technologies' in the percolation model, however, the bit strings can represent solutions to actual problems of combinatorial optimisation from operational research and computer science, although the most popular choice of problem is an abstract one, namely Kauffman's NK fitness landscapes. Heuristics for solving real-world design problems are also present in genetic programming and Arthur and Polak's (2006) model of technological evolution, in which technologies are encoded as combinations of index numbers rather than in binary. In all these search models what connects one combination to another are the heuristic processes of exploration. What makes a new combination more valuable than the previous is determined by a function defined outside the model and unchanging during the simulation run, such as the *NK* fitness landscape in models of organisational learning, and the list of desired logic functions in Arthur and Polak (2006). With different processes and a different evaluation function, different innovations would be generated.

Both the SKIN model (Ahrweiler et al., 2004; Gilbert et al., 2007; Pyka et al., 2007) and Padgett's hypercycles model (Padgett, 1997; Padgett et al., 2003, 2012) represent knowledge as an ability to transform something.

In the SKIN model, possession of a set of kenes and an innovation hypothesis gives one the ability to transform input integers – 'capabilities' and 'abilities' – into a new integer, representing the output product. This transformation is computed as a sum of products, normalised by modulo arithmetic. In an early form, Gilbert's Academic Science Structure model (Gilbert, 1997), kenes were coordinates in a two-dimensional plane and thus users could understand visually the model's workings. The current process connecting input kenes to output employs easily computed mathematical relations to define a technology space of means-end relations, but has no empirical grounding and loses the visualisation of the earlier model. Padgett's model borrows from algorithmic chemistry production rules that are easier to write down and work with than modulo arithmetic, and products and production rules could be given meanings from a real-world case study. However, future extensions of the model will need to permit production rules to relate multiple inputs and outputs, just as the innovation hypotheses in the SKIN model does, and as the technologies do in Arthur and Polak's (2006) model.

Models based around search processes tend to show diminishing returns to search effort as knowledge approaches some peak. Completing the search task may take more computing time than the experimenter bothers to process, such as is likely in Arthur and Polak's model where generating a technology set that met all the desired functions on the list was not achieved during their simulation runs. But in the models of organisational learning it is common, and usually accompanied with a convergence in beliefs of the model organisation's agents, as occurs in Lazer and Friedman's (2007) model. In March's (1991) model the environment can be made to change randomly through the process of 'turbulence', but if workers have converged on a single position, then further search is only possible if some source of novelty exists, such as staff 'turnover' introducing workers with new ideas.

The model of Cowan et al. (2007) uses continuous variables to represent knowledge, which might be thought of as coordinates in a multi-dimensional space, though no attempt is made to depict firms' knowledge visually. Since innovation results in positive quantities being added to these variables' values, the resulting knowledge space is open-ended and innovation may continue indefinitely. With sufficient care from the programmer, a version of Silverberg and Verspagen's percolation grid can also allow indefinite progress (ours does not). But by using continuous variables as quantitative inputs and sums of outputs, the 'knowledge elements' of Cowan et al. bear more resemblance to the concept of 'expertise' in the SKIN model and the numbers of instances of particular types of rule held by firms in the hypercycles model. That is, they influence which dimensions or types of knowledge are focused on during innovation. In the case of Cowan et al. the focus comes in deciding which knowledge variable shall be added to, rather than calculating how much shall be added – the amount of knowledge created is the result of all variables input to the CES production function. The SKIN and hypercycles models employ their quantitative measures to decide which knowledge units are to be used. Lack of use leads to these knowledge units being 'forgotten'. By contrast, in Cowan et al., knowledge values can never decrease, nor do discoveries become forgotten in the percolation model, though this latter phenomenon would be easy to add to the program.

As a heuristic search process similar to ant algorithms, the hypercycles model converges on a single stable system state, albeit a stochastic one. Further dynamics must wait for a later extension to this model. The SKIN model, however, is more open-ended. Its mathematical representation of innovation enables its agents to discover as many products as are allowed by the modulus of the arithmetic, and if these have been forgotten by the current agents, then they may be rediscovered with no sense that what knowledge has existed before is in any way less valuable than what is discovered now. A larger chemistry of rules, involving more products, would allow the hypercycles model to have more complex production chains, but as the experiments show, the difficulty of establishing stochastically stable systems to run these chains undergoes a phase transition.

The application of genetic algorithms to the prisoner's dilemma (Axelrod, 1997a, Chapters 1, 2; Lindgren, 1992) uses bit strings – in this case to encode game strategies – and heuristic search. However, by evaluating these strategies in games against each other, rather than against a fixed set of strategies, the model becomes an open-ended system. No strategy or set of strategies can be expected to remain successful and popular indefinitely. Indeed, the population dynamics show periods of relative stability and instability. Variable instability can also be seen in the obsolescence and replacement events among Arthur and Polak's (2006) technologies, which they found tended towards a scale-free frequency distribution in size. Padgett aims to study the percolation of changes through multiple overlapping systems, though the current hypercycles model will need to be extended to study this (Padgett, 2012; Padgett et al., 2012). Nonetheless, simulations of innovation will need to explore this concept further, since it goes to the heart of what innovation is for. As Schumpeter put it, innovation is necessary to distinguish oneself from one's competitors, to offer something worth paying a premium for (Heilbroner, 2000, pp. 294–295). Once competitors have copied an innovation, prices are driven down and profits disappear. In the language of game theory, too much reproduction of the same strategy makes one too predictable, and predictable agents will be taken advantage of by others. Innovation creates feelings of surprise when it appears and the possibility of innovation forces others to pay attention to one's agency and potential. As is known from studies of the workplace, uncertainty is power (Collins, 1992, p. 83; Wilensky, 1964). In the above models, agents who collaborate on knowledge creation or technology production can have fluctuating relations between collaborators or suppliers and customers, but if the systems are not sufficiently open-ended, the knowledge and market dynamics are threatened with an end themselves.

Several models simulate the emergence of network structures between their agents. This was most explicit in Cowan et al. (2007) where network metrics were calculated for the emergent structures across the parameter space and attempts made to relate these to the expansion in knowledge space. But it is also present in the SKIN model and the hypercycles model. The SKIN model includes both temporary collaborative relations, on which network analysis could be performed, and formalised 'alliances' for innovation and 'networks' for extra production, the sizes of the latter of which show some approximation to a scale-free distribution. The networks of firms left in the hypercycles model, after learning-by-doing and forgetting have redistributed the rule instances and killed off empty firms, still reflect the initial network structure between the firms at the start, namely a regular grid. However, by starting with a complete network, any patterns in the structure at the end of the simulation run could be ascribed solely to the distribution of rule instances and the function of the production runs. In all these models there is scope for future studies relating emergent network structure to initial structures.

By now some ideas seem to be emerging about how elements of the various simulation models might be combined in one future model of innovation. At present the SKIN model offers many features, including collaborations in the search for new knowledge, collaboration in the production of commercial products, and evaluation of both in the form of market pricing and profit making. The choice of collaboration partners could include reputational effects, such as involving structural as well as relational credit, as in Cowan et al. (2007). But the representation of knowledge in the SKIN model should be replaced. Assuming a visualisation such as the two-dimensional planes of the science models was not required, then the replacement could be either a genuine design problem, such as the logic gates of Arthur and Polak (2006), or a version of algorithmic chemistry, an extension of that used in Padgett's hypercycles model. Like Silverberg and Verspagen (2005, 2007) there should be some definition of innovation size, so that stylised fact of a scale-free distribution can be sought. Like Arthur and Polak (2006), attention should be paid to the distribution of obsolescence and replacement. Like Cowan et al. (2007), there should be an attempt to address issues of the emergent structures of collaboration and social capital, by correlating knowledge possession and generation to network positions. How changes percolate through overlapping networks should be a long-term aim of study.

One component missing from these models is what economists call 'the demand side'. Although model agents supply components to be used with other agents, whether in knowledge search or economic production, and in that sense are supplying demands for these, in none of these models is the external consumer simulated. For example, there is constant external demand for products in both the SKIN model and Padgett's hypercycles models. Real-world production systems supply to people outside the system. In addition, the producers consist of human beings with many needs that a modeller is likely to omit, such as their needs for food, drink and shelter. Yet production systems for one type of product can often be affected by the availability of other products, such as when rising food prices lead to demands from workers for higher wages, and disruption when these are not forthcoming. Epstein and Axtell's (1996) *Sugarscape*, an early contribution to the field of agent-based computational economics, did model agents with needs for a representative resource, 'sugar', which they sought out and died from lack of. But such considerations are absent from the models in this chapter. Models that fit innovation into a wider economic and social context must wait for future study.

8. Conclusions

HOW WE SIMULATED INNOVATION

Models of Innovation Diffusion, Generation and the Impact of Diffusion

This book has now surveyed various simulation models of innovation generation, diffusion and the impact of diffusion. In this chapter we summarise what was covered, what the simulation models could offer to innovation studies that other aids to thinking had perhaps neglected and what has been learned about social simulation as a research method.

Models of the diffusion of innovations first appeared in Chapter 2, including the epidemic model, which focuses on innovations spreading via social interactions, the probit model, which focuses on heterogeneous agents making adoption decisions in a changing environment, the stock model, in which new adoption is a response to the current level of adoption, and the evolutionary model, in which competing innovations owe their relative adoption success to extrinsic, environmental factors applying selection pressures. The following chapters focused on the epidemic models. Chapter 3 examined the role played by the structure of the social network constraining interactions, including its effects on diffusion reach and speed, and the path-dependent outcomes when two innovations diffused in competition. In these network models agents could be heterogeneous in the attributes of their positions in the social network. Chapter 3 also discussed the information cascades model, in which agents based their adoption decisions on previous ones. Chapter 4 included March's (1991) model of organisational learning, in which the diffusion of knowledge occurred via the medium of a collective, organisational code, and Lazer and Friedman's (2007) model of collective learning, in which network structure had the potential to moderate the rate at which agents explored new solutions to problems rather than exploiting existing ones. In Chapter 5, epidemic diffusion was moderated by a preference for similarity in content in Axelrod's model of cultural influence (Axelrod, 1997a, b), Gilbert's (1997) academic science structure model, and our own simulation of scientific publication and citation (Watts and Gilbert, 2011).

The science models represented the diffusion of ideas between academic papers, as may be traced by the references to past papers and by common authorship. Social interaction was again the basic of diffusion in our model in Chapter 6, but this time agents in heterogeneous contexts could reach different adoption decisions given the same diffusing innovation. Environment-based interdependencies between technological innovations constrained which innovations went well together for which agents. Evolutionary fitness pressures also drove innovation dynamics in some of the models of Chapter 7. The determination of fitness could be exogenous to the model, such as in Arthur and Polak's (2006) model of technological evolution, or largely endogenous, as was the case when a population of strategies for the prisoner's dilemma evolved by playing one against another. So, although the epidemic model's focus on social interaction remains the most popular basis for modelling the diffusion of innovations, models can also include what resembles the probit model's focus on agent decision makers' responses to their heterogeneous environments and the evolutionary model's selection pressures on competing innovations.

The generation of innovations was a theme from Chapter 4 onwards. Primarily this was represented as the generation of new combinations of existing ideas, beliefs, practices or technologies, whether in the models of organisational learning of Chapter 4 (Lazer and Friedman, 2007; March, 1991) or our science model (Watts and Gilbert, 2011) in Chapter 5. New combinations were created as part of heuristic search methods, including trial-and-error experimentation, and collaborative co-construction of new combinations by sampling a mixture of beliefs from two or more agents, or the contents of two or more academic papers. The model of Chapter 6 also introduced innovation in context, as a diffusing technology could acquire new value and meaning simply by being adopted in a new environment or by a new type of agent. Chapter 7 introduced models in which new structures emerged, including the chains of technologies in Silverberg and Verspagen's (2005, 2007) percolation model, the social networks of Cowan et al. (2007) and the hypercycle systems of economic production (Padgett, 1997; Padgett et al., 2003; Padgett et al., 2012). So, although it could be said that innovations in each model were generated from the re-combination of pre-existing parts, what was meant by 'parts' and 'innovations' varied.

The impact of innovation diffusion appeared first in Chapter 3 where one diffusing innovation had the power to inhibit the diffusion of another. The potential for markets to become locked into inferior technologies was discussed in this regard. In the information cascades model, diffusion was also given a purpose: knowledge of some external environment. Chapter 4 developed this idea with models in which agents were able collectively to obtain knowledge, or relatively fit solutions to problems, that they were unlikely to achieve acting in isolation from each other. The role of diffusion in obtaining knowledge was also present in Chapters 5, 6 and 7. However, diffusion could sometimes fail to produce knowledge. Chapter 3 introduced the problems of groupthink, fads and herd behaviour via the information cascades model. Agents influenced by each other could become locked into consensus positions long after those positions have become erroneous. Chapter 4 introduced the problem of premature convergence, when agents copy each other's beliefs too quickly and converge on a solution that is not the best available. Our science model in Chapter 5 (Watts and Gilbert, 2011) also investigated the scope for this problem in the context of the academic publication practices that lead to clustering and cumulative advantage. Chapter 6 stressed the role that one diffusing innovation can play in catalysing or constraining the adoption of another. In Chapter 7 models of technological evolution included the possibility of one technology's emergence rendering multiple other, interdependent technologies unfit or obsolete, the phenomenon Schumpeter called 'creative destruction'. The impact of diffusion could then either be evaluated in terms of factors exogenous to the system, such as optimisation with regards to a fitness landscape or meeting a list of technological needs, or endogenous factors, such as affecting the diffusion of other technologies.

The models surveyed in this book have thus addressed innovation diffusion, generation and the impact of diffusion, and many models have included more than one of these.

Complex Adaptive Systems and the Emergence of Novelty

As promised in Chapter 1, simulating innovation involved the concept of a complex adaptive system. Most of the models include forms of complexity, with multiple parts, multiple attributes and multiple attribute states, and networks of interdependencies between the parts. They also involved forms of adaptation, such as the processes of trial-and-error experimentation and learning-from-others in the models of organisational learning in Chapter 4 and the model of technology adoption in Chapter 6, the collective learning through peer review and the methods used for sampling co-authors and references for papers in the science model of Watts and Gilbert (2011) in Chapter 5, and the learning-by-doing in Padgett's hypercycles model of economic production in Chapter 7 (Padgett et al., 2003). Like ants in a colony, relatively simple agents were able to adapt collectively to develop new structures or patterns that were in some sense more stable, persistent or fit than that which existed before, and the agents were able to achieve this without intervention or foresight from a user or programmer outside the system. The learning and innovating systems self organised.

That they were able to do so stems from a number of facts: that there were relatively stable or fit system states to find, that there were relatively few of them, and that there were relatively short paths to them through the vast number of possible system states. That is, the system state spaces were searchable. As Kauffman argued with respect to autocatalytic sets and the origins of order (Kauffman, 1993, 1996), the existence of these ordered states stems from the mathematics of combinations. Given sufficient numbers of diverse parts, and sufficient numbers of interdependency relations, catalysis and constraints between them, it becomes inevitable that the possible ways of being for those parts will include ones where the parts form self-maintaining structures. This is Kauffman's 'order for free' (Kauffman, 1996). However, as demonstrated by Kauffman's own NK fitness landscapes (see Chapter 4) and also by K-Sat constraint satisfaction problems (Chapter 6), the existence of ordered states does not mean a system will self-organise to a particularly good one. The difficulty of solving these two model systems can be tuned by the adjustment of one or two parameters. Likewise, Padgett's hypercycles model (Chapter 7) had a parameter, 'rule-complexity', that is, the number of types of product related by production rules, which controlled the chances of self-maintaining production systems emerging. The simulation models make apparent that a period of technological evolution or progress reflects the existence of a technology space searchable by the particular methods performed by society's scientists, engineers and other innovators. We might not know the structure of our own, real-world technology and knowledge spaces – Simon's (1962) suggestion that they involve modularity and decomposability into subtasks remains a good one – but the dynamics of technological and scientific development over the past few centuries implies that the spaces have been, up to now, structured in such ways that they are searchable.

WHAT SIMULATION BRINGS TO INNOVATION STUDIES

Generative Social Science

Improvement from search, or optimisation, is just one example of a system-level pattern emerging from the interactions of component parts. Previous chapters include several other examples of micro-level mechanisms generating macro-level patterns. There were the S-shaped adoption and bell-shaped adoption rate curves of Chapter 2, emergent network structures (Chapters 3 and 7), emergent consensus, fads, belief convergence and paper topic clustering (Chapters 3, 4 and 5), scale-free authorship and citation frequency distributions from Chapter 5's science models, path dependent outcomes (Chapters 3 and 6), emergent hypercycles of Chapter 7 and scale-free frequency distributions of sizes of technological change events (Chapter 7). As was noted in Chapter 2, there are multiple modelling approaches for generating S-shaped and bell-shaped curves, but otherwise no models other than agent-based ones are available for explaining how these patterns might be generated. Once idealised but unrealistic assumptions have been dropped, such as agent homogeneity or uniform chances of social mixing, in favour of agents heterogeneous in their attributes, luck, environments and social network positions, then agent-based modelling becomes almost the only modelling approach capable of representing the combinations of diversity and interdependency from which novel patterns can emerge.

This generative social science, centred on agent-based simulation modelling, connects with empirical research at both micro and macro levels. Inspiration for the choice of micro-level behaviour or mechanisms came from social psychologists' and sociologists' principles, such as homophily, a preference for interacting with those similar to us, and also from ethnographic studies, observations of how innovation occurs in organisations (Chapter 4), among academic scientists (Chapter 5) and in technological projects (Chapter 6). Evidence for macro-level patterns included historians' accounts of technologies and projects in various industries (Chapters 2, 6, 7), social network data (Chapter 3) and bibliometric data (Chapter 5). Patent data would have been another possible source.

Whereas other modelling approaches might have had as their aim either deduction, i.e. inference to what must be, given various assumptions and observations, or induction, i.e. inference to what will probably be, given patterns in past data, this generative social science is abductive. Given observed patterns we infer mechanisms that might plausibly explain these patterns. We make no claims for these being the true mechanisms – in many cases there may be other plausible mechanisms, and insufficient data to decide between rival explanations, as was the case in Chapter 2 for the diffusion models intended to explain S-shaped adoption curves. In many cases, however, particular mechanisms leave empirical traces other than the macro-level patterns they generate. Observing a pattern and knowing from simulation studies at least one mechanism that can generate that pattern can be the cause for further studies, such as experiments on behaviour in the psychology lab and ethnographic studies in the workplace. So identifying a plausible generative mechanism through a simulation model is not the end of research, but rather a step to further research.

Rethinking Innovation

A benefit of having explanations from simulation models for the patterns
seen in innovation studies is that we can conceive of and examine alternative explanations. For instance, we are freed from the so-called 'linear
model', which split innovation into distinct, neat phases, in particular
distinguishing the origin of an innovation (invention, discovery, product
launch) from its diffusion or adoption. Simulation models can represent
composite innovations capable of adaptation by users (Chapter 6), or the
result of collective co-construction by groups of people (Chapters 4, 5).
We are freed from assuming that there must be one single point of origin
for an innovation. We are also freed from postulates concerning the scale
of talent and effort on the part of an originator. Studies of path dependence, positive feedback loops, fads, clustering and cumulative advantage
(Chapters 2, 3, 5, 6) indicate that successful diffusion need not imply any
exceptional talent or effort on the part of the originators. If we reward
people for the success of the innovations they have played a part in developing, it is not clear that there will be repeat performances when it comes
to their next innovation. In addition, diffusion success is no guarantee of
superior quality on the part of an innovative idea, practice or technology.
That we study the things we do to the extent we do may stem from social
reinforcing loops building on historical accidents, not the underlying
structure of material reality reflected by science.

We are also freed from various simplifying assumptions that other
modelling approaches required in order to be tractable. We neither need
to work with one typical agent, nor with a homogeneous population of
agents. Agents and diffusing technologies do not need to be independent
of each other. Networks of social interactions and other interdependencies can be structured, rather than complete or with perfectly even mixing.
Agent decision makers can be bounded rational. They need neither perfect
knowledge nor infinite speed to process it. They can employ heuristic
methods, both as individuals and as a collective, and what constitutes the
most rational way to reach an adoption decision or solve a combinatorial
problem may vary with the problem or task environment and the means
by which rewards are distributed to agents.

Prediction and Policy

In contrast to other forms of modelling, especially statistical modelling,
we did not attempt to make forecasts with these simulation models. In
Chapter 2 we saw how random variability made it very difficult to make
useful forecasts of diffusion over time. While an innovation was still taking

off, forecasts of later adoption were unreliable, and by the time the eventual adoption could be forecast reliably, there was little time left in which to react. This problem was identified in a very simple one-innovation, homogeneous-agents epidemic model. Once the diffusion of multiple interdependent technologies among agents in heterogeneous contexts was considered (Chapter 6), forecasting adoption success seemed even harder. In path-dependent systems, simple differences in the order in which particular events occurred could make big differences to the eventual outcome (Chapters 3, 6). But there is still the possibility of some systems being predictable. Both fitness landscapes (Chapter 4) and constraint satisfaction problems (Chapter 6) can sometimes have a relatively small number of (locally) optimal solutions, as determined by the parameters, and in some cases the relative value of the solution may correspond to the number of paths leading to that solution via heuristic search. As with other types of complex adaptive system, attractors in the system state space may be few in number and identifiable from the dynamics. In such cases, it would be theoretically possible to state the probabilities of particular simulation outcomes. Whether a real-world example of a landscape or a constraint system could be identified and predicted, however, remains to be seen. As elsewhere in the social sciences, predictions made by the members of a system have the potential to change that system, and either self-fulfil or self-refute the prediction.

If the emergence, diffusion success and diffusion impact of a particular innovation cannot be forecast from simulation models, what can they offer to policy makers? In many models factors were identified that affected the simulation outcomes. Outcomes included diffusion success, collective intelligence, organisation learning and technological progress. The factors included rates of performance of heuristic search methods, such as learning from others, network structure, and the method for distributing opportunities to construct new solutions. These are factors that in organisations and academic sciences can certainly be influenced by policy makers, though the influence is not always as direct as intended. For example, a manager can issue a new directive concerning who should speak with whom within the firm, but they are unlikely to police that directive outside company buildings and working hours, and probably do not monitor conversations within the firm either. However, for most models where we studied learning or problem-solving performance, we found more than one factor could influence the balance between exploration and exploitation. What the ideal rate of learning from others was, for example, would depend on the current social network structure. Managers considering whether to alter one of these factors, such as by restructuring the organisation or relocating some of its workgroups, lack accurate information concerning what social network structure they currently have (informal communications are particularly hard to monitor), how often people actually interact and how much they learn from each other when they do. In addition, managers do not know the structure of the tasks faced by their organisations. They know there is likely to be a network structure or an interaction rate with peak optimising performance for a task. They do not know, however, whether that peak requires more or less interaction from their employees, and more network links or fewer, than the current level. Simulation models of organisational learning can show us performance variation for a toy problem, such as NK fitness landscapes, but performance variation can be expected to be different for real-world problems. Even if we felt confident that some organisational change would improve the average level of learning in the organisation, this might not be what a manager wanted to achieve. As argued in Chapter 4, sometimes the rewards go not for average performance but for occasional, lucky peaks in performance – sometimes the winner takes all – and so both competing organisations and competing employees may last longer in business if they take risks and vary their performance more. Making policy recommendations on the basis of simulation models can seem rather too much like the advice given in management gurus' business manuals. Pronouncements can be offered, such as: 'Innovation comes from interaction. Interact more to be more innovative!' The advice can be backed up with references to case studies – particular organisations or particular simulation models where this appeared to apply. But there is no guarantee that the advice can be acted upon in another organisation, or that it will have similar success if it is.

FUTURE DIRECTIONS

Empirical Generalisations

This indicates one challenge for future simulation research: to improve the connection with empirical cases. Models such as the ones in this book have demonstrated the concepts of complex adaptive systems and generative mechanisms, and we should now be able to postulate their presence when we observe real-world cases of macro-level patterns, such as scale-free distributions or clustered social networks. In addition, when we think some real-world agents are fulfilling the steps of some mechanism, we now know what type of patterns this will generate. Thus the toy models have expanded our imaginations and analytical capabilities. But most of the real-world systems we research are likely to involve multiple processes, with some being implemented more strongly than others, and some processes interfering with the effects of others. In Chapter 5 we mentioned the attempt in Watts and Gilbert (2011) to combine several generative mechanisms in one model of academic publication. The model was able to fit several empirical patterns, including geometric growth curves and scale-free frequency distributions of papers per author and citations per paper. But once we added processes to select papers and authors for the value of their contents, fitting those distributions seemed to become very much harder. So combining mechanisms while trying still to generate plausible output is a non-trivial problem.

The complexity of real-world social systems might lead one to doubt whether they are amenable to theorising and modelling at all. General models, applicable with minor tuning to many different cases, may not be available. Simulation modellers seeking to understand a particular case may have to develop a particular simulation model for that system. Social systems may seem to lack the equivalent of the general laws employed in physics.

Empirical evidence does exist, however, for generalisations holding despite the variety of contexts, including several generalisations related to innovation. For social simulation, in common with other social research approaches, these generalisations call for both investigations into their implications and explanations of their emergence. Explaining patterns by identifying possible generative mechanisms is, as highlighted in this book, a particular strength of agent-based modelling. Future social simulation researchers might like to address the following empirical generalisations about qualitative and quantitative innovation.

First, there is the growth in qualitative innovation. The number of types of good or service available to a person has increased from an estimated few hundred about 10 000 years ago, to at least the 10 to the power of 10 items in a US city today for which a distinct barcode is provided (Beinhocker, 2007, pp. 456–457). Villani et al. (2007) suggest this increase be explained as a process of exaptive bootstrapping. Derived from discussions of biological evolution (Gould, 2002; Gould and Vrba, 1982), exaptation is the phenomenon of a thing adapted to one purpose or environment happening by chance to be capable of fitting some new purpose or environment, and thereafter being improved in that new dimension of fitness. For example, birds with feathered wings evolved because such wings help to regulate temperature. Accidentally, the wings also help with flight, and birds came to be naturally selected on that basis instead. Put another way, objects are capable of multiple interpretations, and shifts in the dominant interpretation can lead to improvements being sought in new directions. As these improvements are being made, new capacities for re-interpretation can emerge. Hence reinterpreted objects lead on to more reinterpreted objects. These dynamics of re-interpretation, a boot-strapping process, are postulated to explain not only the growth in the number of things, but also the difficulty in foreseeing the emergence of a particular, important innovation when it still lies beyond a couple of re-interpretations in the future. Villani et al. (2007) describe an agent-based model of exaptation, involving changes in what functionality supplier and customer agents attribute to objects. Simulating innovation, then, involves simulating not just evolution in the objects' attributes but also in agents' cognitions about objects. One avenue for future research will be to under-stand further the processes by which so many new directions in which to develop are generated and recognised.

A second type of generalisation concerns quantitative innovation, the process by which things improve quantitatively in some quality (become better, faster, cheaper, etc.) The best known of these is Moore's Law, which holds that the number of transistors on integrated circuits doubles approximately every 2 years. But this is just one of many generalisations concerning the rate of improvement in quality that held or are holding for various technologies over periods of decades (Lienhard, 1979, 1985, 2006). This is despite wars and economic cycles affecting investment levels in R&D and despite each technology's development involving multiple countries. The growth can continue even when, as in the case of the cost of computing power (Nordhaus, 2007), the underlying basic component technology changed more than once (from mechanical, to relays and vacuum tubes, to transistors, to microprocessors) and applications and markets changed also (from mostly military or academic uses, to the era of the personal computer and on into the Internet age). There have also been general trends in the general trends about improvements in quality (Lienhard, 1979, 1985, 2006). Figure 8.1 employs the data in Lienhard (2006, p. 129). Lienhard's definition of quality varies with technology – for example, speed for land, water and air vehicles, and power-to-weight ratio for their engines – and the reader is directed to his writings for more details on his calculations and justification for them. Here we just use his data as an illustration of the possibility of quantitative generalisations. Until around 1841 a technology beginning a trajectory in quality improvements could be expected to take around 41 years to double in quality, roughly the length of a person's working life. From the 1840s on, the length of time required for a doubling in quality in a newly emerged technology decreased over time at an exponential rate. Generative mechanisms are required to explain not only the stability of the individual quality growth rates, but also the stability of the pre- and post-1841 phases of quality growth rate dynamics, as well as for the 1841 phase transition. The

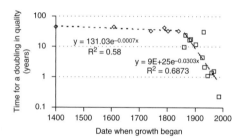

Figure 8.1   *An empirical generalisation about empirical generalisations about quantitative innovation. Lienhard's (2006, p. 129) data on rates of growth in the qualities of various technologies. For technologies whose improvement began before the 1840s, quality took about 41 years to double. Since the 1840s the quality-doubling time has shrunk exponentially.*

suggestions in Lienhard (2006) include new educational institutions and the professionalisation of research in the early nineteenth century, and the role of people's expectations for growth in motivating quantitative innovation, so future social simulation models of innovation may have to incorporate some of these.

Turning to a third type of generalisation about innovation, complexity science researchers have identified a number of scaling laws for cities and business organisations, laws of the form 'the more something is x, the more/less it is y' (Bettencourt, Lobo, Helbing et al., 2007; Bettencourt, Lobo and Strumsky, 2007; Bettencourt et al., 2010; Bettencourt, Lobo and West, 2008). For example, the bigger the city in terms of population size, the lower the number of petrol stations per head of population. Resource efficiency improves with city size, as it does with the size (in body mass) of biological systems (Bettencourt, Lobo, Helbing, et al., 2007). However, indicators of innovation, including the number of patents per capita and the number of 'supercreatives' and entrepreneurs, increase with city size (Bettencourt, Lobo and Strumsky, 2007). Business corporations, on the other hand, become less productive of innovations as they get bigger. This may be connected to an important difference between cities and businesses in terms of expected lifespan. Cities last for centuries or even millennia. Business corporations, if they survive their initial year or two, can be expected to last decades. The empirical evidence for the urban scaling laws is excellent in terms of correlations, and applies across different countries and cultures. But an explanation is lacking for the correlations, a causal mechanism in terms of micro-level behaviour that could explain the emergence and persistence of these macro-level patterns. One was given for scaling laws for biological systems (Enquist, Brown and West, 1998; West, Brown, and Enquist, 1997, 1999), in terms of resource networks in organisms, such as in the lungs and in the blood circulation system. We know both cities and businesses contain networks for resource distribution (e.g. roads and corridors) and social interaction. But we lack a theory to tell us how these are formed, where and when they are and how this determines resource use and innovation production. Agent-based models are, of course, the obvious choice for testing theories about generative mechanisms. With now more than half the world's population living in cities, and future scarcity of several currently important resources, the development of such a theory is particularly desirable. The chapters in Lane (2009) represent early steps towards this end.

The development of simulation models to represent either cities and organisations in general, or specific case studies, may well require something we described in Chapter 7 as completing the loop. That is, the development of models of innovation generation and diffusion in which innovation has an impact on the people doing the generating and diffusing. There is a field that studies artificial societies – Epstein and Axtell's (1996) famous Sugarscape model was an early example – and agent-based models of economic systems do exist (Delli Gatti, 2008; LeBaron and Tesfatsion, 2008; Tesfatsion, 2002), even if they are still rare. But collectively the models in this book have shown innovations have both social causes and social effects, and in future there will hopefully be more attempts to incorporate both in the same model. As we suggested at the beginning of Chapter 1, in the aftermath of a financial crisis that has to date had immense economic, social and political impact, and in which innovation played an important part, we need more tools to place innovation in wider socio-economic contexts.

Model Replication and Teaching

What part may this book play in the future development of simulations of innovation? One contribution is pedagogic. We have gathered together in one place references to different models of diffusion, organisational learning, science, knowledge dynamics and technological evolution, and highlighted some of their similarities and differences, strengths and limitations. This will be convenient for those trying to teach these models to others. Perhaps more importantly, it will help in the communication of the concepts illustrated by these models, especially the generative mechanisms, such as heuristic search, homophily and cumulative advantage. Another contribution is to the research fields in which these models were developed: innovation studies, science and technology studies, organisation studies, etc. Several of these models are classics, well-cited in academic literature, but not always replicated by someone other than the original authors. Repeatability is one of the most highly regarded features of experiments and observations in other sciences, but much rarer in social sciences. When one has reproduced someone else's results, one is both more ready to trust those results, and also more confident that one has understood the process that produced them. Except for our own, nearly every model discussed in this book was successfully replicated in either NetLogo or Excel/VBA. In addition, many of the study results reported in the original papers and books were reproduced with our versions. At no point did we need to copy a chart from another paper or book. Such success in replication shows that social simulation can aim at this standard, as found in other sciences. By collecting our models together in one place and posting them online, free to download and run, we hope we have facilitated the reader's own understanding of both innovation models and the development of simulation programs.

Appendix

All the models developed or reproduced by the authors are available on a website.

http://www.simian.ac.uk/resources/models/simulating-innovation

In some browsers it may be necessary to right-click on a filename and select an option called something like 'Save target as . . .'.

EXCEL FILES

The .xls files were developed using Excel 2003 in Windows XP and tested in Excel 2010 in Windows 7, and some use Visual Basic for Applications (VBA) macros. Users of corporate or university computers may need to request special permission from their IT administrator to run VBA macros.

NETLOGO FILES

The .nlogo files have been tested with NetLogo 5.0.4. NetLogo currently works under Windows, Mac and Linux operating systems. NetLogo is free to download from:

http://ccl.northwestern.edu/netlogo/

Users of corporate or university computers may need their IT administrator to install it. Non-commercial uses of NetLogo are free. For other uses, consult the NetLogo website for their terms and conditions of use.

OUR TERMS AND CONDITIONS OF USE

If you use these programs in your work, please cite the book as well as giving the URL for the website from which you downloaded them and the date on which you downloaded them.

A suitable form of citation is:

Watts, Christopher and Gilbert, Nigel (2014) *Simulating Innovation: Computer-based Tools for Rethinking Innovation.* Cheltenham, UK and Northampton, MA, USA: Edward Elgar.

Watts, Christopher (2013) SI_Model_Comparison.xls. Retrieved 24 April 2013 from http://www.simian.ac.uk/resources/models/simulating-innovation.

The authors have tried to remove all bugs and flaws from these programs, but readers download and run the programs at their own risk.

Readers are welcome to modify the programs providing that it is clearly documented within the program that they have made changes and what those changes are, and our own copyright messages are left in place. The programs are made available under the GNU Public Licence. Readers wishing to make commercial use of the programs should contact us to discuss terms.

CONTACTING THE AUTHORS

We hope these programs will continue to be useful for years to come, but the complexity of today's IT systems and the high rate of innovation in this area mean that we cannot guarantee this.

Feedback, including reports of possible bugs and suggestions for improvement, are welcome, but the authors cannot guarantee they will be able to respond to every message received. In so far as time permits us, we will try to correct any faults and update the versions on the website.

We had a lot of fun developing these, and we wish readers lots of fun in using them!

Christopher Watts and Nigel Gilbert

References

Abernathy, W. J. (1978). *The Productivity Dilemma: Roadblock to Innovation in the Automobile Industry*. Baltimore, MD: Johns Hopkins University Press.

Abernathy, W. J. and Clark, K. B. (1985). Innovation: mapping the winds of creative destruction. *Research Policy*, **14**(1), 3–22.

Acemoglu, D., Dahleh, M. A., Lobel, I. and Ozdaglar, A. (2011). Bayesian learning in social networks. *The Review of Economic Studies*, **78**(4), 1201–1236.

Ackermann, F. and Eden, C. (2011). *Making Strategy: Mapping Out Strategic Success* (2nd edn). London: SAGE.

Ackoff, R. L. (1981). The art and science of mess management. *Interfaces*, **11**(1), 20–26.

Ahrweiler, P. and Gilbert, N. (2005). Caffe Nero: the evaluation of social simulation. *Journal of Artificial Societies and Social Simulation*, **8**(4), 14.

Ahrweiler, P., Pyka, A. and Gilbert, N. (2004). Simulating knowledge dynamics in innovation networks (SKIN). In R. Leombruni and M. Richiardi (eds), *Industry and Labor Dynamics: The Agent-based Computational Economics Approach: Proceedings of the Wild@ace2003 Workshop*, Torino, Italy, 3–4 October 2003 (pp. 284–296). Hackensack, NJ: World Scientific.

Akerlof, G. A. and Shiller, R. J. (2009). *Animal Spirits: How Human Psychology Drives the Economy, and Why it Matters for Global Capitalism*. Princeton, NJ: Princeton University Press.

Akrich, M. (1992). The description of technical objects. In W. E. Bijker and J. Law (eds), *Shaping Technology/Building Society: Studies in Sociotechnical Change*. Cambridge, Mass.; London: MIT Press, pp. 205–224.

Akrich, M., Callon, M. and Latour, B. (2002). The key to success in innovation part I: the art of interessement. *International Journal of Innovation Management*, **6**(2), 187.

Akrich, M., Callon, M., Latour, B. and Monaghan, A. (2002). The key to success in innovation part II: the art of choosing good spokespersons. *International Journal of Innovation Management*, **6**(2), 207.

Albert, R. and Barabási, A.-L. (2002). Statistical mechanics of complex networks. *Reviews of Modern Physics*, **74**(1), 47–97.

Altenberg, L. (1997). NK fitness landscapes. In T. Back, D. Fogel and Z. Michalewicz (eds), *Handbook of Evolutionary Computation*. Oxford: Oxford University Press, pp. B2.7:5–B2.7:10.

Argote, L. and Greve, H. R. (2007). A behavioral theory of the firm: 40 years and counting. Introduction and impact. *Organization Science*, **18**(3), 337–349.

Arthur, W. B. (1989). Competing technologies, increasing returns, and lock-in by historical events. *The Economic Journal*, **99**(394), 116–131.

Arthur, W. B. (1994). *Increasing Returns and Path Dependence in the Economy*. Ann Arbor, MI: University of Michigan Press.

Arthur, W. B. (2010). *The Nature of Technology: What it is and How it Evolves*. London: Penguin.

Arthur, W. B., Ermoliev, Y. M. and Kaniovski, Y. M. (1986). Strong laws for a class of path-dependent stochastic-processes with applications. *Lecture Notes in Control and Information Sciences*, **81**, 287–300.

Arthur, W. B., Ermoliev, Y. M. and Kaniovski, Y. M. (1987). Path-dependent processes and the emergence of macro-structure. *European Journal of Operational Research*, **30**(3), 294–303.

Arthur, W. B. and Polak, W. (2006). The evolution of technology within a simple computer model. *Complexity*, **11**(5), 23–31.

Ashby, W. R. (1956). *An Introduction to Cybernetics*. London: Chapman & Hall.

Ashby, W. R. (1958). Requisite variety and its implications for the control of complex systems. *Cybernetica*, **1**(2), 83–99.

Axelrod, R. M. (1997a). *The Complexity of Cooperation: Agent-based Models of Competition and Collaboration*. Princeton, NJ: Princeton University Press.

Axelrod, R. M. (1997b). The dissemination of culture: a model with local convergence and global polarization. *Journal of Conflict Resolution*, **41**(2), 203–226.

Axtell, R., Axelrod, R., Epstein, J. and Cohen, M. (1996). Aligning simulation models: A case study and results. *Computational & Mathematical Organization Theory*, **1**(2), 123–141.

Bagni, R., Berchi, R. and Cariello, P. (2002). A comparison of simulation models applied to epidemics. *Journal of Artificial Societies and Social Simulation*, **5**(3).

Bak, P. (1997). *How Nature Works: The Science of Self-organized Criticality*. Oxford: Oxford University Press.

Bak, P., Tang, C. and Wiesenfeld, K. (1987). Self-organized criticality: an explanation of 1/F noise. *Physical Review Letters*, **59**(4), 381–384.

Balconi, M., Brusoni, S. and Orsenigo, L. (2010). In defence of the linear model: an essay. *Research Policy*, **39**(1), 1–13.

Banerjee, A. V. (1992). A simple model of herd behavior. *The Quarterly Journal of Economics*, **107**(3), 797–817.

Barabási, A.-L. (2002). *Linked: The New Science of Networks*. Cambridge, MA: Perseus Publishing.

Barabási, A.-L. and Albert, R. (1999). Emergence of scaling in random networks. *Science*, **286**(5439), 509–512.

Bass, F. M. (1969). A new product growth model for consumer durables. *Management Science*, **15**(5), 215–227.

Bedau, M. A., McCaskill, J. S., Packard, N. H. and Rasmussen, S. (2010). Living technology: exploiting life's principles in technology. *Artificial Life*, **16**(1), 89–97.

Beer, S. (1959). *Cybernetics and Management*. New York: Wiley.

Beinhocker, E. D. (2007). *The Origin of Wealth: Evolution, Complexity, and the Radical Remaking of Economics*. London: Random House Business.

Bentley, R. A., Ormerod, P. and Batty, M. (2011). Evolving social influence in large populations. *Behavioral Ecology and Sociobiology*, **65**(3), 537–546.

Berkes, F., Colding, J. and Folke, C. (2003). *Navigating Social-ecological Systems: Building Resilience for Complexity and Change*. Cambridge: Cambridge University Press.

Bernard, H. R. and Killworth, P. D. (1977). Informant accuracy in social network data II. *Human Communications Research*, **4**(1), 3–18.

Bernard, H. R., Killworth, P. D., Evans, M. J., McCarty, C. and Shelley, G. A. (1988). Studying social relations cross-culturally. *Ethnology*, **27**(2), 155–179.

Bertalanffy, L. v. (1971). *General System Theory: Foundations, Development, Applications*. London: Allen Lane.

Bettencourt, L. M. A., Lobo, J., Helbing, D., Kuhnert, C. and West, G. B. (2007). Growth, innovation, scaling, and the pace of life in cities. *Proceedings of the National Academy of Sciences of the United States of America*, **104**(17), 7301–7306.

Bettencourt, L. M. A., Lobo, J. and Strumsky, D. (2007). Invention in the city: Increasing returns to patenting as a scaling function of metropolitan size. *Research Policy*, **36**(1), 107–120.

Bettencourt, L. M. A., Lobo, J., Strumsky, D. and West, G. B. (2010). Urban scaling and its deviations: revealing the structure of wealth, innovation and crime across cities. *Plos One*, **5**(11).

Bettencourt, L. M. A., Lobo, J. and West, G. B. (2008). Why are large cities faster? Universal scaling and self-similarity in urban organization and dynamics. *European Physical Journal B*, **63**(3), 285–293.

Bigbee, A., Cioffi-Revilla, C. and Luke, S. (2007). Replication of Sugarscape using MASON. In T. Terano, H. Kita, H. Deguchi and K. Kijima (eds), *Agent-Based Approaches in Economic and Social Complex Systems IV*, Vol. 3. Berlin: Springer, pp. 183–190.

Bijker, W. E. (1992). The social construction of fluorescent lighting, or how an artefact was invented during its diffusion stage. In W. E. Bijker and J. Law (eds), *Shaping Technology/Building Society: Studies in Sociotechnical Change*. Cambridge, MA: MIT Press, pp. 75–102.

Bijker, W. E. (1995). *Of Bicycles, Bakelites, and Bulbs: Toward a Theory of Sociotechnical Change*. Cambridge, MA and London: MIT Press.

Bijker, W. E., Hughes, T. P. and Pinch, T. J. (1987). *The Social Construction of Technological Systems: New Directions in the Sociology and History of Technology*. Cambridge, MA and London: MIT Press.

Bijker, W. E. and Law, J. (1992). *Shaping Technology/Building Society: Studies in Sociotechnical Change*. Cambridge, MA and London: MIT Press.

Bikhchandani, S., Hirshleifer, D. and Welch, I. (1992). A theory of fads, fashion, custom, and cultural change as informational cascades. *Journal of Political Economy*, **100**(5), 992–1026.

Bikhchandani, S., Hirshleifer, D. and Welch, I. (1998). Learning from the behavior of others: conformity, fads, and informational cascades. *Journal of Economic Perspectives*, **12**(3), 151–170.

Blanchard, O. (2012). *In the Wake of the Crisis: Leading Economists Reassess Economic Policy*. Cambridge, MA: MIT Press.

Boerner, K., Klavans, R., Patek, M., Zoss, A. M., Biberstine, J. R., Light, R. P., Larivière, V. and Boyack, K. W. (2012). Design and update of a classification system: the UCSD Map of Science. *Plos One*, **7**(7).

Boerner, K., Maru, J. T. and Goldstone, R. L. (2004). The simultaneous evolution of author and paper networks. *Proceedings of the National Academy of Sciences of the United States of America*, **101**(Suppl. 1), 5266–5273.

Boorman, S. A. and White, H. C. (1976). Social structure from multiple networks. II. Role structures. *American Journal of Sociology*, **81**(6), 1384–1446.

Bradford, S. C. (1985). Sources of information on specific subjects (reprinted from *Engineering: An illustrated weekly journal*, vol 137, pg 85–86, 1934). *Journal of Information Science*, **10**(4), 176–180.

Breiger, R. L., Carley, K. M. and Pattison, P. (2003). Dynamic social network modeling and analysis: workshop summary and papers. Washington DC: National Academies Press.

Bronk, R. (2009). *The Romantic Economist: Imagination in Economics*. Cambridge: Cambridge University Press.

Brooks, F. P. (1975). *The Mythical Man-month: Essays on Software Engineering*. Reading, MA: Addison-Wesley Pub. Co.

Brown, J. S. and Duguid, P. (1991). Organizational learning and communities-of-practice: toward a unified view of working, learning, and innovating. *Organization Science*, **2**(1), 40–57.

Brown, J. S. and Duguid, P. (2000). *The Social Life of Information*. Boston, MA: Harvard Business School Press.

Brown, J. S. and Duguid, P. (2001). Knowledge and organization: a social-practice perspective. *Organization Science*, **12**(2), 198–213.

Buchanan, M. (2002). *Small World: Uncovering Nature's Hidden Networks*. London: Weidenfeld & Nicolson.

Burt, R. S. (1992). *Structural Holes: The Social Structure of Competition*. Cambridge, MA: Harvard University Press.

Burt, R. S. (2007). *Brokerage and Closure: An Introduction to Social Capital*. Oxford: Oxford University Press.

Burton, R. E. and Kebler, R. W. (1960). The 'half-life' of some scientific and technical literatures. *American Documentation*, **11**(1–4), 18–22.

Caldarelli, G. (2007). *Scale-free Networks: Complex Webs in Nature and Technology*. Oxford: Oxford University Press.

Caldart, A. A. and Oliveira, F. (2010). Analysing industry profitability: a 'complexity as cause' perspective. *European Management Journal*, **28**(2), 95–107.

Carrington, P. J. and Scott, J. (2011). *The SAGE Handbook of Social Network Analysis*. London: SAGE.

Cartwright, D. and Harary, F. (1956). Structural balance: a generalization of Heider theory. *Psychological Review*, **63**(5), 277–293.

Castellano, C., Marsili, M. and Vespignani, A. (2000). Nonequilibrium phase transition in a model for social influence. *Physical Review Letters*, **85**(16), 3536–3539.

Centola, D., Eguiluz, V. M. and Macy, M. W. (2007). Cascade dynamics of complex propagation. *Physica A: Statistical Mechanics and Its Applications*, **374**(1), 449–456.

Centola, D. and Macy, M. (2007). Complex contagions and the weakness of long ties. *American Journal of Sociology*, **113**(3), 702–734.

Chandler, A. D., Hikino, T. and Von Nordenflycht, A. (2005). *Inventing the Electronic Century: The Epic Story of the Consumer Electronics and Computer Industries*. Cambridge, MA and London: Harvard University Press.

Checkland, P. (1998). *Systems Thinking, Systems Practice*. Chichester, UK: Wiley.

Christensen, C. M. (1997). *The Innovator's Dilemma: When New*

Technologies Cause Great Firms to Fail. Boston, MA: Harvard Business School Press.

Clark, A. (1997). *Being There: Putting Brain, Body, and World Together Again*. Cambridge, MA and London: MIT Press.

Clerc, M. (2006). *Particle Swarm Optimization*. London: ISTE.

Clerc, M. and Kennedy, J. F. (2002). The particle swarm: explosion, stability, and convergence in a multidimensional complex space. *IEEE Transactions on Evolutionary Computation*, **6**(1), 58–73.

Cohen, B. (1997). *The Edge of Chaos: Financial Booms, Bubbles, Crashes and Chaos*. Chichester, UK: Wiley.

Cohen, M. D., Burkhart, R., Dosi, G., Egidi, M., Marengo, L., Warglien, M. and Winter, S. (1996). Routines and other recurring action patterns of organizations: contemporary research issues. *Industrial & Corporate Change*, **5**(3), 653–688.

Cole, P. F. (1962). New look at reference scattering. *Journal of Documentation*, **18**(2), 58–64.

Coleman, J. S. (1988). Social capital in the creation of human capital. *American Journal of Sociology*, **94**, S95–S120.

Coleman, J. S. (1990). *Foundations of Social Theory*. Cambridge, MA: Belknap Press of Harvard University Press.

Collins, R. (1992). *Sociological Insight: An Introduction to Non-obvious Sociology*, 2nd edn. Oxford: Oxford University Press.

Collins, R. (1998). *The Sociology of Philosophies: A Global Theory of Intellectual Change*. Cambridge, MA: Belknap Press of Harvard University Press.

Collins, R. (2004). *Interaction Ritual Chains*. Princeton, NJ: Princeton University Press.

Conner, D. (1998). *Leading at the Edge of Chaos: How to Create the Nimble Organization*. Chichester, UK: John Wiley.

Cooper, R. G. (1990). Stage-gate systems: a new tool for managing new products. *Business Horizons*, **33**(3), 44–54.

Corne, D., Dorigo, M. and Glover, F. (1999). *New Ideas in Optimization*. London: McGraw-Hill.

Cowan, R., Jonard, N. and Zimmermann, J. B. (2007). Bilateral collaboration and the emergence of innovation networks. *Management Science*, **53**(7), 1051–1067.

Creswell, J. W. and Plano Clark, V. L. (2007). *Designing and Conducting Mixed Methods Research*. Thousand Oaks, CA: London: SAGE.

Cross, R. L. and Parker, A. (2004). *The Hidden Power of Social Networks: Understanding How Work Really Gets Done in Organizations*. Boston, MA: Harvard Business School Press.

Cyert, R. M., March, J. G. and Clarkson, G. P. E. (1964). *A Behavioral Theory of the Firm*. Englewood Cliffs, NJ: Prentice Hall.

D'Adderio, L. (2008). The performativity of routines: theorising the influence of artefacts and distributed agencies on routines dynamics. *Research Policy*, **37**(5), 769–789.

David, P. A. (1985). Clio and the economics of QWERTY. *American Economic Review*, **75**(2), 332.

David, P. A. (2001). Path dependence, its critics and the quest for 'historical economics'. In P. Garrouste & S. Ioannides (Eds.), *Evolution and path dependence in economic ideas: past and present*. Cheltenham, UK and Northampton, MA: Edward Elgar Publishing, pp. 15–41.

Davies, S. (1979). *The Diffusion of Process Innovations*. Cambridge: Cambridge University Press.

Davis, J. P., Eisenhardt, K. M. and Bingham, C. B. (2007). Developing theory through simulation methods. *Academy of Management Review*, **32**(2), 480–499.

Deffuant, G., Amblard, F., Weisbuch, G. and Faure, T. (2002). How can extremism prevail? A study based on the relative agreement interaction model. *Journal of Artificial Societies and Social Simulation*, **5**(4).

Delli Gatti, D. (2008). *Emergent Macroeconomics: An Agent-based Approach to Business Fluctuations*. Milan: Springer Verlag.

Dittrich, P. and Banzhaf, W. (1998). Self-evolution in a constructive binary string system. *Artificial Life*, **4**(2), 203–220.

Dittrich, P., Ziegler, J. and Banzhaf, W. (2001). Artificial chemistries: a review. *Artificial Life*, **7**(3), 225–275.

Dopson, S. and Fitzgerald, L. (2005). *Knowledge to Action? Evidence-based Health Care in Context*. Oxford: Oxford University Press.

Dorigo, M. and Blum, C. (2005). Ant colony optimization theory: a survey. *Theoretical Computer Science*, **344**(2–3), 243–278.

Dorigo, M., Maniezzo, V. and Colorni, A. (1996). Ant system: optimization by a colony of cooperating agents. *IEEE Transactions on Systems Man and Cybernetics Part B–Cybernetics*, **26**(1), 29–41.

Dorogovtsev, S. N. and Mendes, J. F. F. (2003). *Evolution of Networks: From Biological Nets to the Internet and WWW*. Oxford: Oxford University Press.

Dreu, C. K. W. d. and Vries, N. K. d. (2001). *Group Consensus and Minority Influence: Implications for Innovation*. Oxford: Blackwell.

Drossel, B. and Schwabl, F. (1992). Self-organized criticality in a forest-fire model. *Physica A*, **191**(1–4), 47–50.

Drossel, B. and Schwabl, F. (1993). Self-organization in a forest-fire model. *Fractals: Complex Geometry Patterns and Scaling in Nature and Society*, **1**(4), 1022–1029.

Dunbar, R. I. M. (1992). Neocortex size as a constraint on group-size in primates. *Journal of Human Evolution*, **22**(6), 469–493.

Dunbar, R. I. M. (1996). *Grooming, Gossip and the Evolution of Language*. London: Faber and Faber.

Dunham, J. B. (2005). An agent-based spatially explicit epidemiological model in MASON. *Journal of Artificial Societies and Social Simulation*, **9**(1).

Eden, C. (1988). Cognitive mapping. *European Journal of Operational Research*, **36**(1), 1–13.

Edmonds, B., Gilbert, N., Ahrweiler, P. and Scharnhorst, A. (2011). Simulating the Social Processes of Science. *Journal of Artificial Societies and Social Simulation*, **14**(4).

Edmonds, B. and Hales, D. (2003). Replication, replication and replication: Some hard lessons from model alignment. *Journal of Artificial Societies and Social Simulation*, **6**(4).

Eigen, M. (1971). Selforganization of matter and evolution of biological macromolecules. *Naturwissenschaften*, **58**(10), 465–523.

Eigen, M. (1979). *The Hypercycle: A Principle of Natural Self-Organization*. Berlin: Springer.

Eigen, M. and Schuster, P. (1977). Hypercycle: principle of natural self-organization. A. Emergence of hypercycle. *Naturwissenschaften*, **64**(11), 541–565.

Eigen, M. and Schuster, P. (1978a). Hypercycle: principle of natural self-organization. B. Abstract hypercycle. *Naturwissenschaften*, **65**(1), 7–41.

Eigen, M. and Schuster, P. (1978b). Hypercycle: principle of natural self-organization. C. Realistic hypercycle. *Naturwissenschaften*, **65**(7), 341–369.

Enquist, B. J., Brown, J. H. and West, G. B. (1998). Allometric scaling of plant energetics and population density. *Nature*, **395**(6698), 163–165.

Epstein, J. M. and Axtell, R. (1996). *Growing artificial societies: social science from the bottom up*. Washington DC and London: Brookings Institution Press and MIT Press.

Erdős, P. and Rényi, A. (1959). On random graphs I. *Publicationes Mathematicae*, **6**, 290–297.

Fagerberg, J., Mowery, D. C. and Nelson, R. R. (2005). *The Oxford Handbook of Innovation*. Oxford: Oxford University Press.

Feldman, M. S. (2000). Organizational routines as a source of continuous change. *Organization Science*, **11**(6), 611–629.

Feldman, M. S. (2003). A performative perspective on stability and change in organizational routines. *Industrial and Corporate Change*, **12**(4), 727–752.

Feldman, M. S. and Pentland, B. T. (2003). Reconceptualizing organizational routines as a source of flexibility and change. *Administrative Science Quarterly*, **48**(1), 94–118.

Feldman, M. S. and Rafaeli, A. (2002). Organizational routines as sources of connections and understandings. *Journal of Management Studies*, **39**(3), 309–331.

Ferlie, E., Fitzgerald, L., Wood, M. and Hawkins, C. (2005). The non-spread of innovations: the mediating role of professionals. *Academy of Management Journal*, **48**(1), 117–134.

Fleming, L., Mingo, S. and Chen, D. (2007). Collaborative broker-age, generative creativity, and creative success. *Administrative Science Quarterly*, **52**(3), 443–475.

Fleming, L. and Sorenson, O. (2001). Technology as a complex adaptive system: evidence from patent data. *Research Policy*, **30**(7), 1019–1039.

Fontana, W. (1992). *Algorithmic Chemistry*, Vol. 10. Reading: Addison-Wesley Publ Co.

Fontana, W. (2006). The topology of the possible. In A. Wimmer and R. Kössler (eds), *Understanding Change: Models, Methodologies, and Metaphors*. Houndmills, UK: Palgrave Macmillan, pp. 67–84.

Forrester, J. W. (1961). *Industrial Dynamics*. Cambridge, MA: MIT Press.

Frank, R. H. and Cook, P. J. (2010). *The Winner-take-all Society: Why the Few at the Top Get So Much More than the Rest of Us*. London: Virgin Books.

Frenken, K. (2001). Fitness landscapes, heuristics and technological para-digms: a critique on random search models in evolutionary econom-ics. In D. M. Dubois (ed.), *Computing Anticipatory Systems*,Vol. 573. Melville, NY: American Institute of Physics, pp. 558–565.

Frenken, K. (2006a). *Innovation, Evolution and Complexity Theory*. Cheltenham, UK and Northampton, MA: Edward Elgar Publishing.

Frenken, K. (2006b). Technological innovation and complexity theory. *Economics of Innovation and New Technology*, **15**(2), 137–155.

Frey, B. S. (2008). *Happiness: A Revolution in Economics*. Cambridge, MA and London: MIT.

Frydman, R. and Goldberg, M. D. (2011). *Beyond Mechanical Markets: Asset Price Swings, Risk, and the Role of the State*. Princeton, NJ: Princeton University Press.

Fuller, S. (2000). *The Governance of Science: Ideology and the Future of the Open Society*. Buckingham, UK: Open University Press.

Fuller, S. (2004). In search of vehicles for knowledge governance: on the need for institutions that creatively destroy social capital. In N. Stehr (ed.), *The Governance of Knowledge*. New Brunswick, NJ: Transaction pp. 41–76.

Galam, S. (2004). Sociophysics: a personal testimony. *Physica A: Statistical Mechanics and Its Applications*, **336**(1–2), 49–55.

Gardner, M. (1970). Mathematical games: the fantastic combinations of John Conway's new solitaire game 'life'. *Scientific American*, **223**(October), 120–123.

Gavetti, G. (2005). Cognition and hierarchy: rethinking the microfoundations of capabilities' development. *Organization Science*, **16**(6), 599–617.

Gavetti, G. and Levinthal, D. (2000). Looking forward and looking backward: cognitive and experiential search. *Administrative Science Quarterly*, **45**(1), 113–137.

Gavetti, G., Levinthal, D. and Ocasio, W. (2007). Neo-carnegie: the Carnegie school's past, present, and reconstructing for the future. *Organization Science*, **18**(3), 523–536.

Gavetti, G., Levinthal, D. A. and Rivkin, J. W. (2005). Strategy making in novel and complex worlds: the power of analogy. *Strategic Management Journal*, **26**(8), 691–712.

Geroski, P. A. (2000). Models of technology diffusion. *Research Policy*, **29**(4–5), 603–625.

Gigerenzer, G., Todd, P. M. and ABC Research Group. (1999). *Simple Heuristics That Make Us Smart*. Oxford: Oxford University Press.

Gilbert, N. (1977). Referencing as persuasion. *Social Studies of Science*, **7**(1), 113–122.

Gilbert, N. (1997). A simulation of the structure of academic science. *Sociological Research Online*, **2**(2).

Gilbert, N. (2008). *Agent-based Models*. Los Angeles and London: SAGE.

Gilbert, N., Ahrweiler, P. and Pyka, A. (2007). Learning in innovation networks: some simulation experiments. *Physica A: Statistical Mechanics and Its Applications*, **378**(1), 100–109.

Gilbert, N. and Troitzsch, K. G. (2005). *Simulation for the Social Scientist*, 2nd edn. Maidenhead, UK: Open University Press.

Gillies, D. (2007). Lessons from the history and philosophy of science for research assessment systems. In A. O'Hear (ed.), *Philosophy of Science*. Cambridge: Cambridge University Press, pp. vii, 278.

Gillies, D. (2010). Lessons from the history and philosophy of science for research assessment systems. *Journal of Biological Physics and Chemistry*, **10**(4), 158–164.

Gladwell, M. (2000). *The Tipping Point: How Little Things Can Make a Big Difference*. London: Little, Brown and Company.

Gleick, J. (1987). *Chaos: Making A New Science*. New York: Viking.

Glover, F. (1989). Tabu Search. Part I. *ORSA Journal on Computing*, **1**(3), 190.

Glover, F. (1990). Tabu Search. Part II. *ORSA Journal on Computing*, **2**(1), 4.

Godin, B. (2006). The linear model of innovation: the historical construction of an analytical framework. *Science, Technology & Human Values*, **31**(6), 639–667.

Goldenberg, J., Libai, B., Muller, E. and Stremersch, S. (2010). The evolving social network of marketing scholars. *Marketing Science*, **29**(3), 561–567.

Goldstein, J., Hazy, J. K. and Lichtenstein, B. B. (2010). *Complexity and the nexus of leadership: leveraging nonlinear science to create ecologies of innovation*. New York: Palgrave Macmillan.

Gomes, C. P. and Selman, B. (2005). Computational science: can get satisfaction. *Nature*, **435**(7043), 751–752.

Gould, S. J. (2002). *The Structure of Evolutionary Theory*. Cambridge, MA: Belknap.

Gould, S. J. and Vrba, E. S. (1982). Exaptation: a missing term in the science of form. *Paleobiology*, **8**(1), 4–15.

Granovetter, M. S. (1973). The strength of weak ties. *American Journal of Sociology*, **78**(6), 1360–1380.

Griliches, Z. (1957). Hybrid corn: an exploration in the economics of technological change. *Econometrica*, **25**(4), 501–522.

Grimm, V., Revilla, E., Berger, U., Jeltsch, F., Mooij, W. M., Railsback, S. F., Thulke, H.-H., Weiner, J., Wiegand, T. and DeAngelis, D. L. (2005). Pattern-oriented modeling of agent-based complex systems: lessons from ecology. *Science*, **310**(5750), 987–991.

Hales, D., Rouchier, J. and Edmonds, B. (2003). Model-to-model analysis. *Journal of Artificial Societies and Social Simulation*, **6**(4).

Hamill, L. and Gilbert, N. (2009). Social circles: a simple structure for agent-based social network models. *Journal of Artificial Societies and Social Simulation*, **12**(2).

Harhoff, D., Narin, F., Scherer, F. M. and Vopel, K. (1999). Citation frequency and the value of patented inventions. *Review of Economics and Statistics*, **81**(3), 511–515.

Hedström, P. and Bearman, P. (2009). *The Oxford Handbook of Analytical Sociology*. Oxford: Oxford University Press.

Hegselmann, R. and Krause, U. (2002). Opinion dynamics and bounded confidence: models, analysis and simulation. *Journal of Artificial Societies and Social Simulation*, **5**(3).

Hegselmann, R. and Krause, U. (2006). Truth and cognitive division of labour first steps towards a computer aided social epistemology. *Journal of Artificial Societies and Social Simulation*, **9**(3).

Heilbroner, R. L. (2000). *The Worldly Philosophers: The Lives, Times, and Ideas of the Great Economic Thinkers* (revised 7th edn). London: Penguin.

Henderson, R. M. and Clark, K. B. (1990). Architectural innovation: the reconfiguration of existing product technologies and the failure of established firms. *Administrative Science Quarterly*, **35**(1), 9–30.

Hernando, A., Villuendas, D., Vesperinas, C., Abad, M. and Plastino, A. (2010). Unravelling the size distribution of social groups with information theory in complex networks. *European Physical Journal B*, **76**(1), 87–97.

Hjorland, B. and Nicolaisen, J. (2005). Bradford's law of scattering: ambiguities in the concept of 'subject'. In F. Crestani and I. Ruthven (eds), *Context: Nature, Impact, and Role*. Berlin: Springer-Verlag Berlin, pp. 96–106.

Hofbauer, J. and Sigmund, K. (1988). *The Theory of Evolution and Dynamical Systems: Mathematical Aspects of Selection*. Cambridge: Cambridge University Press.

Hofbauer, J. and Sigmund, K. (1998). *Evolutionary Games and Population Dynamics*. Cambridge: Cambridge University Press.

Holland, J. H. (1975). *Adaptation in Natural and Artificial Systems: An Introductory Analysis with Applications to Biology, Control, and Artificial Intelligence*. Ann Arbor, MI: University of Michigan Press.

Homans, G. C. (1951). *The Human Group*. London: Routledge & Kegan Paul.

Horgan, J. (1995). From complexity to perplexity. *Scientific American*, **272**(6), 104–109.

Hutchins, E. (1995a). *Cognition in the Wild*. Cambridge, MA and London: MIT Press.

Hutchins, E. (1995b). How a cockpit remembers its speeds. *Cognitive Science*, **19**(3), 265–288.

Irving, R. W. (1985). An efficient algorithm for the "stable roommates" problem. *Journal of Algorithms*, **6**(4), 577–595.

Jin, Y. and Branke, H. (2005). Evolutionary optimization in uncertain environments: a survey. *IEEE Transactions on Evolutionary Computation*, **9**(3), 303–317.

Kahneman, D. (2011). *Thinking, Fast and Slow*. London: Allen Lane.

Kahneman, D., Slovic, P. and Tversky, A. (1982). *Judgment under Uncertainty: Heuristics and Biases*. Cambridge: Cambridge University Press.

Kauffman, S. A. (1993). *The Origins of Order: Self-organization and Selection in Evolution*. Oxford: Oxford University Press.

Kauffman, S. A. (1995). *At Home in the Universe: The Search for Laws of Self-Organization and Complexity*. Oxford: Oxford University Press.

Kauffman, S. A. (1996). *At Home in the Universe: The Search for Laws of Self-Organization and Complexity.* London: Penguin.

Kauffman, S. A. (2000). *Investigations.* Oxford: Oxford University Press.

Keen, S. (2011). *Debunking Economics: The Naked Emperor Dethroned?* (Rev. and expanded edn). London: Zed.

Kermack, W. O. and McKendrick, A. G. (1927). A Contribution to the mathematical theory of epidemics. *Proceedings of the Royal Society of London. Series A, Containing Papers of a Mathematical and Physical Character*, **115**(772), 700–721.

Killworth, P. D. and Bernard, H. R. (1976). Informant accuracy in social network data. *Human Organization*, **35**(3), 269–286.

Killworth, P. D. and Bernard, H. R. (1978). The reversal small-world experiment. *Social Networks*, **1**, 159–192.

Killworth, P. D., McCarty, C., Johnsen, E. C., Bernard, H. R. and Shelley, G. A. (2006). Investigating the variation of personal network size under unknown error conditions. *Sociological Methods & Research*, **35**(1), 84–112.

Kirkpatrick, S., Gelatt, C. D. and Vecchi, M. P. (1983). Optimization by simulated annealing. *Science*, **220**(4598), 671–680.

Kirkpatrick, S. and Selman, B. (1994). Critical behavior in the satisfiability of random boolean expressions. *Science*, **264**(5163), 1297–1301.

Klein, G. A. (1998). *Sources of Power: How People Make Decisions.* Cambridge, MA and London: MIT Press.

Klein, G. A. (2009). *Streetlights and Shadows: Searching for the Keys to Adaptive Decision Making.* Cambridge, MA: MIT Press.

Klemm, K., Eguiluz, V. M., Toral, R. and Miguel, M. S. (2003). Global culture: a noise-induced transition in finite systems. *Physical Review E*, **67**(4).

Klemm, K., Eguiluz, V. M., Toral, R. and San Miguel, M. (2003). Role of dimensionality in Axelrod's model for the dissemination of culture. *Physica A:Statistical Mechanics and Its Applications*, **327**(1–2), 1–5.

Koza, J. R. (1992). *Genetic Programming: On the Programming of Computers by Means of Natural Selection.* Cambridge, MA: MIT Press.

Koza, J. R. (1994). *Genetic Programming II: Automatic Discovery of Reusable Programs.* Cambridge, MA: MIT Press.

Koza, J. R. (1996). *Genetic Programming: Proceedings of the First Annual Conference, 1996.* Cambridge, MA: MIT Press.

Koza, J. R. (1999). *Genetic Programming III: Darwinian Invention and Problem Solving.* San Francisco, CA: Morgan Kaufmann.

Koza, J. R. (2003). *Genetic Programming IV: Routine Human–Competitive Machine Intelligence.* Norwell, MA: Kluwer Academic Publishers.

Koza, J. R., Andre, D., Bennett III, F. H. and Keane, M. A. (1996). Use of automatically defined functions and architecture-altering operations in automated circuit synthesis using genetic programming. In J. R. Koza (ed.), *Genetic Programming: Proceedings of the First Annual Conference, 1996*. Cambridge, MA: MIT Press, pp. 132–140.

Krugman, P. R. (1991). *Geography and Trade*. Leuven, Belgium: Leuven University Press.

Kuhn, T. S. (1962). *The Structure of Scientific Revolutions*. Chicago, IL: University of Chicago Press.

Kurtz, C. F. and Snowden, D. J. (2003). The new dynamics of strategy: sense-making in a complex and complicated world. *IBM Systems Journal*, **42**(3), 462–483.

Lane, D. A. (2009). *Complexity Perspectives in Innovation and Social Change*. Dordrecht, The Netherlands: Springer.

Langton, C. G. (1995). *Artificial Life: An Overview*. Cambridge, MA: MIT Press.

Latour, B. (1996). *Aramis: Or the Love of Technology*. Cambridge, MA: Harvard University Press.

Law, A. M. (2006). *Simulation modeling and Analysis*, 4th edn. Boston and London: McGraw-Hill.

Law, J. and Callon, M. (1992). The life and death of an aircraft: a network analysis of technical change. In W. E. Bijker and J. Law (Eds.), *Shaping technology/building society: studies in sociotechnical change*. Cambridge, MA: MIT Press, pp. 21–52.

Layard, R. (2011). *Happiness: Lessons from a New Science* (New fully rev. and updated edn). London: Penguin Books.

Lazarsfeld, P. F. and Merton, R. K. (1954). Friendship as a social process: a substantive and methodological analysis. In M. Berger, T. Abel and C. H. Page (eds), *Freedom and Control in Modern Society: Written in Honour of R. M. MacIver*. New York: Van Nostrand, pp. 18–66.

Lazer, D. and Friedman, A. (2007). The network structure of exploration and exploitation. *Administrative Science Quarterly*, **52**(4), 667–694.

LeBaron, B. and Tesfatsion, L. (2008). Modeling macroeconomies as open-ended dynamic systems of interacting agents. *American Economic Review*, **98**(2), 246–250.

Levinthal, D. A. (1997). Adaptation on rugged landscapes. *Management Science*, **43**(7), 934–950.

Levitt, B. and March, J. G. (1988). Organizational learning. *Annual Review of Sociology*, **14**(1), 319–338.

Lewin, R. (1993). *Complexity: Life at the Edge of Chaos*. London: Dent.

Lienhard, J. H. (1979). The rate of technological improvement before and after the 1830s. *Technology and Culture*, **20**(3), 515–530.

Lienhard, J. H. (1985). Some ideas about growth and quality in technology. *Technological Forecasting and Social Change*, **27**(2–3), 265–281.

Lienhard, J. H. (2006). *How Invention Begins: Echoes of Old Voices in the Rise of New Machines*. Oxford: Oxford University Press.

Lindgren, K. (1992). *Evolutionary Phenomena in Simple Dynamics*, Vol. 10. Reading, UK: Addison-Wesley Publ. Co.

Lloyd, S. (2001). Measures of complexity: a nonexhaustive list. *Control Systems, IEEE*, **21**(4), 7–8.

Lorrain, F. and White, H. C. (1971). Structural equivalence of individuals in social networks. *Journal of Mathematical Sociology*, **1**(1), 49–80.

Lotka, A. J. (1926). The frequency distribution of scientific productivity. *Journal of the Washington Academy of Sciences*, **16**(12), 317–324.

Luhmann, N. (1990). *Essays on Self-reference*. New York: Columbia University Press.

MacKenzie, D. (2009). All those arrows. [Review of the book *Fool's Gold: How Unrestrained Greed Corrupted a Dream, Shattered Global Markets and Unleashed a Catastrophe*]. *London Review of Books*, **31**(12), 20–22.

MacKenzie, D. (2011a). The credit crisis as a problem in the sociology of knowledge. *American Journal of Sociology*, **116**(6), 1778–1841.

MacKenzie, D. (2011b). How to make money in microseconds. *London Review of Books*, **33**, 16–18.

Macy, M. and Sato, Y. (2010). The surprising success of a replication that failed. *Journal of Artificial Societies and Social Simulation*, **13**(2).

Malerba, F. and Brusoni, S. (2007). *Perspectives on Innovation*. Cambridge: Cambridge University Press.

March, J. G. (1991). Exploration and exploitation in organizational learning. *Organization Science*, **2**(1), 71–87.

March, J. G. and Simon, H. A. (1958). *Organizations*. New York: Wiley.

Mason, W. and Watts, D. J. (2012). Collaborative learning in networks. *Proceedings of the National Academy of Sciences of the United States of America*, **109**(3), 764–769.

Maturana, H. R. and Varela, F. J. (1980). *Autopoiesis and Cognition: The Recognition of the Living*. Dordrecht, the Netherlands: Reidel.

May, R. M. (2001). *Stability and Complexity in Model Ecosystems*. Princeton, NJ: Princeton University Press.

McCarty, C., Killworth, P. D., Bernard, H. R., Johnsen, E. C. and Shelley, G. A. (2001). Comparing two methods for estimating network size. *Human Organization*, 60(1), 28–39.

McCloskey, D. N. (1985). The loss function has been mislaid: the rhetoric of significance tests. *American Economic Review*, **75**(2), 201.

McCloskey, D. N. and Ziliak, S. T. (1996). The standard error of regressions. *Journal of Economic Literature*, **34**(1), 97–114.

McCloskey, D. N. and Ziliak, S. T. (2009). The unreasonable ineffectiveness of Fisherian 'tests' in biology, and especially in medicine. *Biological Theory*, **4**(1), 44–53.

McLean, P. D. and Padgett, J. F. (1997). Was Florence a perfectly competitive market? Transactional evidence from the Renaissance. *Theory and Society*, **26**(2–3), 209–244.

McPherson, M., Smith-Lovin, L. and Cook, J. M. (2001). Birds of a feather: homophily in social networks. *Annual Review of Sociology*, **27**, 415–444.

Merton, R. K. (1968a). The Matthew Effect in science: the reward and communication systems of science are considered. *Science*, **159**(3810), 56–63.

Merton, R. K. (1968b). *Social Theory and Social Structure* (1968 enl. ed.). New York: Free Press.

Merton, R. K. (1973). *The Sociology of Science: Theoretical and Empirical Investigations*. Chicago, IL: University of Chicago Press.

Merton, R. K. (1988). The Matthew Effect in science, II: cumulative advantage and the symbolism of intellectual property. *Isis*, **79**(4), 606–623.

Merton, R. K. and Barber, E. G. (2004). *The travels and adventures of serendipity: a study in sociological semantics and the sociology of science*. Princeton, NJ: Princeton University Press.

Meyer, M. (2011). Bibliometrics, stylized facts and the way ahead: how to build good social simulation models of science? *Journal of Artificial Societies and Social Simulation*, **14**(4).

Meyer, M., Lorscheid, I. and Troitzsch, K. G. (2009). The development of social simulation as reflected in the first ten years of JASSS: a citation and co-citation analysis. *Journal of Artificial Societies and Social Simulation*, **12**(4), A224–A243.

Meyer, M., Zaggl, M. A. and Carley, K. M. (2011). Measuring CMOT's intellectual structure and its development. *Computational and Mathematical Organization Theory*, **17**(1), 1–34.

Milgram, S. (1967). Small-world problem. *Psychology Today*, **1**(1), 61–67.

Milo, R., Shen-Orr, S., Itzkovitz, S., Kashtan, N., Chklovskii, D. and Alon, U. (2002). Network motifs: simple building blocks of complex networks. *Science*, **298**(5594), 824–827.

Mingers, J. (1995). *Self-producing Systems: Implications and Applications of Autopoiesis*. New York and London: Plenum Press.

Mitchell, M. (1996). *An Introduction to Genetic Algorithms*. Cambridge, MA: MIT.

Monge, P. R. and Contractor, N. S. (2003). *Theories of Communication Networks*. Oxford: Oxford University Press.

Morecroft, J. and Robinson, S. (2006). Comparing discrete-event simulation and system dynamics: modelling a fishery. Paper presented at the Proceedings of the Operational Research Society Simulation Workshop.

Nelson, R. R. and Winter, S. G. (1982). *An Evolutionary Theory of Economic Change*. Cambridge, MA and London: Belknap Press.

Newman, M. E. J. (2003). The structure and function of complex networks. *Siam Review*, **45**(2), 167–256.

Newman, M. E. J. (2010). *Networks: An Introduction*. Oxford: Oxford University Press.

Newman, M. E. J., Barabási, A.-L. and Watts, D. J. (2006). *The Structure and Dynamics of Networks*. Princeton, NJ: Princeton University Press.

Nicolaisen, J. and Hjorland, B. (2007). Practical potentials of Bradford's law: a critical examination of the received view. *Journal of Documentation*, **63**(3), 359–377.

Nordhaus, W. D. (2007). Two centuries of productivity growth in computing. *Journal of Economic History*, **67**(1), 128–159.

Padgett, J. F. (1997). The emergence of simple ecologies of skill: a hypercycle approach to economic organization. In W. B. Arthur, S. N. Durlauf and D. A. Lane (eds), *The Economy as an Evolving Complex System II*. Reading, MA: Advanced Book Program/Perseus Books, pp. xii.

Padgett, J. F. (2001). Organizational genesis, identity and control: the transformation of banking in renaissance Florence. In J. E. Rauch and A. Casella (eds), *Networks and Markets*. New York: Russell Sage Foundation, pp. 211–257.

Padgett, J. F. (2012). From chemical to social networks. In J. F. Padgett and W. W. Powell (Eds.), *The Emergence of Organizations and Markets*. Princeton, NJ: Princeton University Press, pp. 92–114.

Padgett, J. F. and Ansell, C. K. (1993). Robust actions and the rise of the Medici, 1400–1434. *American Journal of Sociology*, **98**(6), 1259.

Padgett, J. F., Lee, D. and Collier, N. (2003). Economic production as chemistry. *Industrial and Corporate Change*, **12**(4), 843–877.

Padgett, J. F., McMahan, P. and Zhong, X. (2012). Economic production as chemistry II. In J. F. Padgett and W. W. Powell (eds), *The Emergence of Organizations and Markets*. Princeton, NJ: Princeton University Press, pp. 70–91.

Padgett, J. F. and Powell, W. W. (2012). *The Emergence of Organizations and Markets*. Princeton, NJ: Princeton University Press.

Pareto, V. (1909 [1972]). *Manual of Political Economy*. London: Macmillan.

Pareto, V. and Bonnet, A. (1909). *Manuel d'économie politique*. Paris: V. Giard & E. Brière.

Pascale, R. T., Millemann, M. and Gioja, L. (2000). *Surfing the Edge of*

Chaos: The Laws of Nature and the New Laws of Business. London: Texere.

Pentland, B. T. and Feldman, M. S. (2005). Organizational routines as a unit of analysis. *Industrial and Corporate Change*, **14**(5), 793–815.

Pentland, B. T. and Feldman, M. S. (2008). Designing routines: on the folly of designing artifacts, while hoping for patterns of action. *Information and Organization*, **18**(4), 235–250.

Pfeffer, J. and Sutton, R. I. (1999). Knowing "what" to do is not enough: Turning knowledge into action (Reprinted from The Knowing-Doing Gap: How Smart Companies Turn Knowledge Into Action). *California Management Review*, 42(1), 83–108.

Pfeffer, J. and Sutton, R. I. (2000). *The Knowing–doing Gap: How Smart Companies Turn Knowledge Into Action*. Boston, MA: Harvard Business School Press.

Pidd, M. (1996). *Tools for Thinking: Modelling in Management Science*. Chichester: Wiley.

Pidd, M. (1999). Just modeling through: a rough guide to modeling. *Interfaces*, **29**(2), 118–132.

Pimm, S. L. (2002). *Food webs*. Chicago, IL: University of Chicago Press.

Poli, R., Langdon, W. B., Marrow, P., Kennedy, J., Clerc, M., Bratton, D. and Holden, N. (2006). Communication, leadership, publicity and group formation in particle swarms. In M. Dorigo, L. M. Gambardella, M. Birattari, A. Martinoli and T. Stutzle (eds), *Ant Colony Optimization and Swarm Intelligence*. Berlin: Springer, pp. 132–143.

Popper, K. R. (1957). *The Poverty of Historicism*. London: Routledge & Paul.

Porter, M. E. (1990). *The Competitive Advantage of Nations*. London: Macmillan.

Price, D. J. d. S. (1963). *Little Science, Big Science*. New York: Columbia University Press.

Price, D. J. d. S. (1965). Networks of scientific papers. *Science*, **149**(3683), 510–515.

Price, D. J. d. S. (1976). A general theory of bibliometric and other cumulative advantage processes. *Journal of the American Society for Information Science*, **27**(5/6), 292–306.

Price, D. J. d. S. (1983). Citation classic: little science, big science. *ISI Current Contents: Social & Behavioral Science*, **29**, 18.

Prigogine, I. and Stengers, I. (1984). *Order Out of Chaos: Man's New Dialogue with Nature*. London: Fontana Paperbacks.

Pujol, J. M., Flache, A., Delgado, J. and Sanguesa, R. (2005). How can social networks ever become complex? Modelling the emergence of complex networks from local social exchanges. *Journal of Artificial Societies and Social Simulation*, **8**(4).

Pyka, A., Gilbert, N. and Ahrweiler, P. (2007). Simulating knowledge-generation and distribution processes in innovation collaborations and networks. *Cybernetics and Systems*, **38**(7), 667–693.

Rahmandad, H. and Sterman, J. (2008). Heterogeneity and network structure in the dynamics of diffusion: comparing agent-based and differential equation models. *Management Science*, **54**(5), 998–1014.

Reinganum, J. F. (1981). Market structure and the diffusion of new technology. *Bell Journal of Economics*, **12**(2), 618–624.

Repenning, N. P. (2002). A simulation-based approach to understanding the dynamics of innovation implementation. *Organization Science*, **13**(2), 109–127.

Rittel, H. W. J. and Webber, M. M. (1973). Dilemmas in a general theory of planning. *Policy Sciences*, **4**(2), 155–169.

Rivkin, J. W. and Siggelkow, N. (2003). Balancing search and stability: Interdependencies among elements of organizational design. *Management Science*, **49**(3), 290–311.

Robertson, D. A. and Caldart, A. A. (2008). Natural science models in management: opportunities and challenges. *Emergence: Complexity & Organization*, **10**(2), 61–75.

Robins, G., Snijders, T., Wang, P., Handcock, M. and Pattison, P. (2007). Recent developments in exponential random graph (p*) models for social networks. *Social Networks*, **29**(2), 192–215.

Robinson, S. (2001). Soft with a hard centre: discrete-event simulation in facilitation. *Journal of the Operational Research Society*, **52**(8), 905–915.

Robinson, S. (2004). *Simulation: The Practice of Model Development and Use*. Chichester: Wiley.

Robinson, S. (2008a). Conceptual modelling for simulation. Part I: definition and requirements. *Journal of the Operational Research Society*, **59**(3), 278–290.

Robinson, S. (2008b). Conceptual modelling for simulation. Part II: a framework for conceptual modelling. *Journal of the Operational Research Society*, **59**(3), 291–304.

Rodan, S. (2005). Exploration and exploitation revisited: extending March's model of mutual learning. *Scandinavian Journal of Management*, **21**(4), 407–428.

Rogers, E. M. (1958). Categorizing the adopters of agricultural practices. *Rural Sociology*, **23**(4), 347–354.

Rogers, E. M. (2003). *Diffusion of Innovations*, 5th edn. New York: Free Press.

Romer, P. M. (1986). Increasing returns and long-run growth. *Journal of Political Economy*, **94**(5), 1002–1037.

Romer, P. M. (1990). Endogenous technological change. *Journal of Political Economy*, **98**(5), S71–S102.

Rosenberg, N. (1982). *Inside the Black Box: Technology and Economics*. Cambridge: Cambridge University Press.

Rosenhead, J. and Mingers, J. (2001). *Rational Analysis for a Problematic World Revisited: Problem Structuring Methods for Complexity, Uncertainty and Conflict* (2nd edn). Chichester, UK: Wiley.

Rouchier, J. (2003). Re-implementation of a multi-agent model aimed at sustaining experimental economic research: the case of simulations with emerging speculation. *Journal of Artificial Societies and Social Simulation*, **6**(4).

Rouchier, J., Cioffi-Revilla, C., Polhill, J. G. and Takadama, K. (2008). Progress in model-to-model analysis. *Journal of Artificial Societies and Social Simulation*, **11**(2).

Ryan, B. and Gross, N. C. (1943). The diffusion of hybrid seed corn in two Iowa communities. *Rural Sociology*, **8**(1), 15–24.

Sandstrom, P. E. (1999). Scholars as subsistence foragers. *Bulletin of the American Society for Information Science and Technology*, **25**(3), 17–20.

Sandstrom, P. E. (2001). Scholarly communication as a socioecological system. *Scientometrics*, **51**(3), 573–605.

Savage, S. L. and Danziger, J. (2012). *The Flaw of Averages: Why We Underestimate Risk in the Face of Uncertainty*. Hoboken, NJ: Wiley.

Savage, S. L., Scholtes, S. and Zweidler, D. (2006). Probability management. *OR/MS Today*, **33**, 21–28.

Sawyer, R. K. (2012). *Explaining Creativity: The Science of Human Innovation* (2nd edn). New York: Oxford University Press.

Scharnhorst, A. (1998). Citation – Networks, science landscapes and evolutionary strategies: comments on theories of citation? *Scientometrics*, **43**(1), 95–106.

Scharnhorst, A. (2001). *Constructing Knowledge Landscapes within the Framework of Geometrically Oriented Evolutionary Theories*. Berlin: Springer-Verlag.

Scharnhorst, A., Boerner, K. and Besselaar, P. V. D. (2012). *Models of Science Dynamics-encounters between Complexity Theory and Information Sciences*. Berlin: Springer.

Schelling, T. C. (1969). Models of segregation. *American Economic Review*, **59**(2), 488.

Schelling, T. C. (1971). Dynamic models of segregation. *Journal of Mathematical Sociology*, **1**(2), 143–186.

Scherer, F. M. (2000). *The Size Distribution of Profits from Innovation.* Norwell, MA: Kluwer Academic Publishers.

Schön, D. A. (1987). *Educating the Reflective Practitioner: Toward a New Design for Teaching and Learning in the Professions.* San Francisco, CA: Jossey-Bass.

Schumpeter, J. A. (1939). *Business Cycles: A Theoretical, Historical, and Statistical Analysis of the Capitalist Process.* New York, London: McGraw-Hill.

Schumpeter, J. A. (1943). *Capitalism, Socialism, and Democracy.* London: G. Allen & Unwin ltd.

Senge, P. M. (1992). *The Fifth Discipline: The Art and Practice of the Learning Organization.* London: Random House Business Books.

Shiffrin, R. M. and Boerner, K. (2004). Mapping knowledge domains. *Proceedings of the National Academy of Sciences of the United States of America*, **101**(Suppl 1), 5183–5185.

Siggelkow, N. and Rivkin, J. W. (2005). Speed and search: designing organizations for turbulence and complexity. *Organization Science*, **16**(2), 101–122.

Silverberg, G. and Verspagen, B. (2005). A percolation model of innovation in complex technology spaces. *Journal of Economic Dynamics & Control*, **29**(1–2), 225–244.

Silverberg, G. and Verspagen, B. (2007). Self-organization of R&D search in complex technology spaces. *Journal of Economic Interaction and Coordination*, **2**(2), 195–210.

Simon, H. A. (1948). *Administrative Behaviour: A Study of the Decision Making Processes in Administrative Organisation.* New York: The Macmillan Co.

Simon, H. A. (1955a). A behavioral model of rational choice. *The Quarterly Journal of Economics*, **69**(1), 99–118.

Simon, H. A. (1955b). On a class of skew distribution functions. *Biometrika*, **42**(3–4), 425–440.

Simon, H. A. (1957). *Models of Man: Social and Rational; Mathematical Essays on Rational Human Behavior in Society Setting.* New York: Wiley.

Simon, H. A. (1962). The architecture of complexity. *Proceedings of the American Philosophical Society*, **106**(6), 467–482.

Simon, H. A. (1991). Bounded rationality and organizational learning *Organization Science*, **2**(1), 125–134.

Simon, H. A. and Newell, A. (1958). Heuristic problem-solving – the next advance in operations-research. *Operations Research*, **6**(1), 1–10.

Slobodkin, L. B. and Rapoport, A. (1974). An optimal strategy of evolution. *The Quarterly Review of Biology*, **49**(3), 181–200.

Small, H. (1973). Co-citation in the scientific literature: a new measure of the relationship between two documents. *Journal of the American Society for Information Science*, **24**(4), 265–269.

Snijders, T. A. B., Pattison, P. E., Robins, G. L. and Handcock, M. S. (2006). New specifications for exponential random graph models. In R. M. Stolzenberg (ed.), *Sociological Methodology* 2006, *Vol* 36. Malden: Wiley-Blackwell, pp. 99–153.

Snijders, T. A. B., van de Bunt, G. G. and Steglich, C. E. G. (2010). Introduction to stochastic actor-based models for network dynamics. *Social Networks*, **32**(1), 44–60.

Snowden, D. and Stanbridge, P. (2004). The landscape of management: Creating the context for understanding social complexity. *Emergence: Complexity & Organization*, **6**(1/2), 140–148.

Snowden, D. J. and Boone, M. E. (2007). A Leaders Framework for Decision Making – Wise executives tailor their approach to fit the complexity of the circumtances they face. *Harvard Business Review*, 85(11), 68–76.

Solé, R. V., Corominas-Murtra, B. and Fortuny, J. (2010). Diversity, competition, extinction: the ecophysics of language change. *Journal of the Royal Society Interface*, **7**(53), 1647–1664.

Solé, R. V., Corominas-Murtra, B., Valverde, S. and Steels, L. (2010). Language networks: their structure, function, and evolution. *Complexity*, **15**(6), 20–26.

Solow, R. M. (1956). A contribution to the theory of economic growth. *Quarterly Journal of Economics*, **70**(1), 65–94.

Sorenson, O., Rivkin, J. W. and Fleming, L. (2006). Complexity, networks and knowledge flow. *Research Policy*, **35**(7), 994–1017.

Soros, G. (1988). *The Alchemy of Finance: Reading the Mind of the Market*. London: Weidenfeld and Nicolson.

Steels, L. (1996). A self-organizing spatial vocabulary. *Artificial Life*, **2**(3), 319–332.

Steels, L. (2002). Grounding symbols through evolutionary language games. In A. Cangelosi and D. Parisi (eds), *Simulating the Evolution of Language*. London: Springer, pp. 211–226.

Steels, L. and Kaplan, F. (1998). Stochasticity as a source of innovation in language games. In C. Adami, R. K. Belew, H. Kitano and C. Taylor (eds), *Artificial Life VI*. Cambridge, MA: MIT Press, pp. 368–376.

Steels, L. and McIntyre, A. (1998). Spatially distributed naming games. *Advances in Complex Systems*, **01**(04), 301–323.

Steglich, C., Snijders, T. A. B. and Pearson, M. (2010). Dynamic networks and behavior: separating selection from influence. In T. F. Liao

(Ed.), *Sociological Methodology*, Vol. 40. Malden: Wiley-Blackwell, pp. 329–393.

Sterman, J. (2000). *Business Dynamics: Systems Thinking and Modeling for a Complex World*. Boston, MA: Irwin McGraw-Hill.

Stoneman, P. (2002). *The Economics of Technological Diffusion*. Oxford, UK and Malden, MA: Blackwell Publishers.

Strogatz, S. H. (2003). *Sync: The Emerging Science of Spontaneous Order*, 1st edn. New York: Theia.

Tarde, G. d. (1899). *Social Laws: An Outline of Sociology*. London: Macmillan & Co.

Tesfatsion, L. (2002). Agent-based computational economics: growing economies from the bottom up. *Artificial Life*, **8**(1), 55–82.

Tett, G. (2009). *Fool's Gold: How Unrestrained Greed Corrupted a Dream, Shattered Global Markets and Unleashed a Catastrophe*. London: Little, Brown.

Thorngate, W., Liu, J. and Chowdhury, W. (2011). The competition for attention and the evolution of science. *Journal of Artificial Societies and Social Simulation*, **14**(4).

Thorngate, W. and Tavakoli, M. (2005). In the long run: biological versus economic rationality. *Simulation & Gaming*, **36**(1), 9–26.

Tocher, K. D. (1963). *The Art of Simulation*. London: English Universities Press.

Trajtenberg, M. (1990). A penny for your quotes: patent citations and the value of innovations. *RAND Journal of Economics*, **21**(1), 172–187.

Turner, A. (2012). *Economics After the Crisis: Objectives and Means*. Cambridge, MA: MIT Press.

Tushman, M. L. and Anderson, P. (1986). Technological discontinuities and organizational environments. *Administrative Science Quarterly*, **31**(3), 439–465.

Villani, M., Bonacini, S., Ferrari, D., Serra, R. and Lane, D. (2007). An agent-based model of exaptive processes. *European Management Review*, **4**(3), 141–151.

Voinov, A. and Bousquet, F. (2010). Modelling with stakeholders. *Environmental Modelling & Software*, **25**(11), 1268–1281.

Von Neumann, J. and Morgenstern, O. (1947). *Theory of Games and Economic Behavior*, 2nd edn. London: Oxford University Press.

Wagner, C. S., Roessner, J. D., Bobb, K., Klein, J. T., Boyack, K. W., Keyton, J., Rafols, I. and Borner, K. (2011). Approaches to understanding and measuring interdisciplinary scientific research (IDR): a review of the literature. *Journal of Informetrics*, **5**(1), 14–26.

Waldrop, M. M. (1993). *Complexity: The Emerging Science at the Edge of Order and Chaos*. London: Viking.

Wasserman, S. and Faust, K. (1994). *Social Network Analysis: metHods and Applications*. Cambridge: Cambridge University Press.

Watts, C. (2012). The impact of firm network structure on the emergence of social structure in Padgett's economic production model. Paper presented at the 8th European Social Simulation Association Conference, Salzburg, Austria.

Watts, C. and Gilbert, N. (2011). Does cumulative advantage affect collective learning in science? An agent-based simulation. *Scientometrics*, **89**(1), 437–463.

Watts, C. J. (2009). *An Agent-based Model of Energy in Social Networks*. PhD thesis, University of Warwick. Retrieved from http://wrap.warwick.ac.uk/2799/.

Watts, D. J. (2003). *Six Degrees: The Science of a Connected Age*. London: Heinemann.

Watts, D. J., Dodds, P. S. and Newman, M. E. J. (2002). Identity and search in social networks. *Science*, **296**(5571), 1302–1305.

Watts, D. J. and Strogatz, S. H. (1998). Collective dynamics of 'small-world' networks. *Nature*, **393**(6684), 440–442.

Weisberg, M. and Muldoon, R. (2009). Epistemic landscapes and the division of cognitive labor. *Philosophy of Science*, **76**(2), 225–252.

Weisbuch, G., Deffuant, G. and Amblard, F. (2005). Persuasion dynamics. *Physica A: Statistical Mechanics and Its Applications*, **353**, 555–575.

Wenger, E. (1998). *Communities of Practice: Learning, Meaning, and Identity*. Cambridge: Cambridge University Press.

West, G. B., Brown, J. H. and Enquist, B. J. (1997). A general model for the origin of allometric scaling laws in biology. *Science*, **276**(5309), 122–126.

West, G. B., Brown, J. H. and Enquist, B. J. (1999). The fourth dimension of life: fractal geometry and allometric scaling of organisms. *Science*, **284**(5420), 1677–1679.

White, H. C., Boorman, S. A. and Breiger, R. L. (1976). Social structure from multiple networks. I. Blockmodels of roles and positions. *American Journal of Sociology*, **81**(4), 730–780.

Whitley, R. (2000). *The Intellectual and Social Organization of the Sciences*, 2nd edn. Oxford: Oxford University Press.

Wiener, N. (1948). *Cybernetics: Or, Control and Communication in the Animal and the Machine*. New York, Paris: Wiley, Hermann et Cie.

Wilensky, H. L. (1964). *The Professionalization of Everyone?* Berkeley, CA: University of California.

Wilensky, U. (1999). NetLogo: Center for Connected Learning and Computer-Based Modeling, Northwestern University, Evanston, IL. Retrieved from http://ccl.northwestern.edu/netlogo/.

Wilensky, U. and Rand, W. (2007). Making models match: replicating an agent-based model. *Journal of Artificial Societies and Social Simulation*, **10**(4).

Wilson, B. (2001). *Soft Systems Methodology: Conceptual Model Building and its Contribution*. Chichester, UK: Wiley.

Winter, S. G., Cattani, G. and Dorsch, A. (2007). The value of moderate obsession: insights from a new model of organizational search. *Organization Science*, **18**(3), 403–419.

Wittgenstein, L. and Anscombe, G. E. M. (1953). *Philosophische Untersuchungen*. Oxford: Blackwell.

Wittgenstein, L., Wright, G. H. v., Rhees, R. and Anscombe, G. E. M. (1978). *Remarks on the Foundations of Mathematics*, 3rd edn. Oxford: Blackwell.

Woolgar, S. (1991). Configuring the user. In J. Law (ed.), *A Sociology of Monsters: Essays on Power, Technology, and Domination*. London: Routledge, pp. 57–99.

Xie, J., Sreenivasan, S., Korniss, G., Zhang, W., Lim, C. and Szymanski, B. K. (2011). Social consensus through the influence of committed minorities. *Physical Review E*, **84**(1).

Zuckerman, H. (1979). *Scientific Elite: Nobel Laureates in the United States*. New York and London: Free Press and Collier Macmillan.

Index

abductive reasoning 190–91
Abernathy, W.J. 198
ability, firms 229–31
absolute advantage 31, 32
academic journals 140, 142–5, 166–9
academic publications
    bibliometric traces of innovations
        136–40
    and clustering among scientists
        145–57
    cumulative advantage 141–5
    cumulative advantage models 157–65
    per author 135, 137–8, 141, 143–5,
        153, 165, 166–7, 170
    towards a combined model of
        science 165–9
Acemoglu, D. 98
Ackerman, F. 15
Ackoff, R.L. 15
ACM_Similarity.nlogo 150
actor-network theory (ANT) 172,
    176–8, 189
adaptation and adoption 18–19
    agent-based simulation 180–87
    constraints as sources of complexity
        176–8
    discussion 187–91
    model of 178–80
    need for new diffusion model 173–4
    simplifying assumptions of
        traditional models 175–6
AdoptAndAdapt.nlogo 180–87
    representations of knowledge,
        technologies, strategies or rules
        232
    representations of recombination
        233
adopter categories 45–51, 175–6
adoption curves 35–6, 38, 43, 46, 51,
    53, 59, 61–3, 135, 150, 175
adoption event simulation 39–41

adoption threshold 52–3, 76
adoption-rate curves 36, 40–41, 46–51,
    61–2, 175
advertising campaigns 36, 38, 63, 66–7,
    72, 79, 169
agent-based simulation (ABS)
    modelling 26–8, 41–3, 49–51
    common applications 27–8
    compared to other simulation
        approaches 44–6
    economic systems 250
    exaptation 248
    knowledge production through
        bilateral cooperation 210–17
    language evolution 134
    model behaviour 183–7
    organisational learning 102–3
    science as search 169
    summary 43–4
    use in adapting/adopting decisions
        180–91
    use for probit models 53–4
Ahrweiler, P. 25, 27, 229, 232, 233, 234
Akerlof, G.A. 4
Akrich, M. 176, 178
Albert, R. 70, 71, 74, 81, 85, 86, 93–4,
    96–7, 141, 195
Algorithmic Chemistry 209, 218, 235,
    237–8
alliance formation 230–31
Altenburg, L. 202
altruistic learning 111–15, 116, 132,
    220, 223–7
analysis of variance (ANOVA) 23
Anderson, P. 198
ant algorithms 220–21
ant-colony optimisation 157–8
anthropology 12, 16
Apple 9–11
appropriateness of innovation 6–7
Aramis transport project 177, 189–90